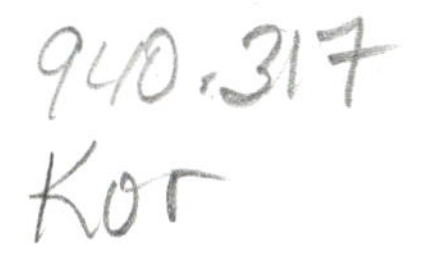

Enemy Aliens, Prisoners of War

McGill-Queen's Studies in Ethnic History
SERIES ONE: Donald Harman Akenson, Editor

1 Irish Migrants in the Canadas
A New Approach
Bruce S. Elliott

2 Critical Years in Immigration
Canada and Australia Compared
Freda Hawkins
(Second edition, 1991)

3 Italians in Toronto
Development of a National Identity, 1875–1935
John E. Zucchi

4 Linguistics and Poetics of Latvian Folk Songs
Essays in Honour of the Sesquicentennial of the Birth of Kr. Barons
Vaira Vikis-Freibergs

5 Johan Schroder's Travels in Canada, 1863
Orm Overland

6 Cass, Ethnicity, and Social Inequality
Christopher McAll

7 The Victorian Interpretation of Racial Conflict
The Maori, the British, and the New Zealand Wars
James Belich

8 White Canada Forever
Popular Attitudes and Public Policy toward Orientals in British Columbia
W. Peter Ward
(Third edition, 2002)

9 The People of Glengarry
Highlanders in Transition, 1745–1820
Marianne McLean

10 Vancouver's Chinatown
Racial Discourse in Canada, 1875–1980
Kay J. Anderson

11 Best Left as Indians
Native-White Relations in the Yukon Territory, 1840–1973
Ken Coates

12 Such Hardworking People
Italian Immigrants in Postwar Toronto
Franca Iacovetta

13 The Little Slaves of the Harp
Italian Child Street Musicians in Nineteenth-Century Paris, London, and New York
John E. Zucchi

14 The Light of Nature and the Law of God
Antislavery in Ontario, 1833–1877
Allen P. Stouffer

15 Drum Songs
Glimpses of Dene History
Kerry Abel

16 Louis Rosenberg
Canada's Jews
(Reprint of 1939 original)
Edited by Morton Weinfeld

17 A New Lease on Life
Landlords, Tenants, and Immigrants in Ireland and Canada
Catharine Anne Wilson

18 In Search of Paradise
The Odyssey of an Italian Family
Susan Gabori

19 Ethnicity in the Mainstream
Three Studies of English Canadian Culture in Ontario
Pauline Greenhill

20 Patriots and Proletarians
The Politicization of Hungarian Immigrants in Canada, 1923–1939
Carmela Patrias

SERIES TWO: John Zucchi, Editor

Enemy Aliens

Prisoners of War

Internment in Canada during the Great War

Bohdan S. Kordan

McGill-Queen's University Press · Montreal & Kingston · London · Ithaca

ISBN 0-7735-2350-2

Legal deposit fourth quarter 2002
Bibliothèque nationale du Québec

Printed in Canada on acid-free paper that is 100% ancient forest free (100% post-consumer recycled), processed chlorine free.

This book has been published with the help of grants from the Publications Fund and the Office of the President of St Thomas More College, the President's Publication Fund of the University of Saskatchewan, and the Chair of Ukrainian Studies Foundation.

McGill-Queen's University Press acknowledges the support of the Canada Council for the Arts for our publishing program. We also acknowledge the financial support of the Government of Canada through the Book Publishing Industry Development Program (BPIDP) for our publsihing activities.

National Library of Canada Cataloguing in Publication

Kordan, Bohdan S
Enemy aliens, prisoners of war : internment in Canada during the Great War / Bohdan S. Kordan.

(McGill-Queen's studies in ethnic history)
Includes bibliographical references and index.
ISBN 0-7735-2350-2

1. World War, 1914–1918—Evacuation of civilians—Canada. 2. World War, 1914–1918—Concentration camps—Canada. 3. Aliens—Canada—History—20th century. I. Title. II. Series.

D627.C2K66 2002 940.3'1771 C2002-903949-5

Frontispiece: To work, Cave and Basin, Banff
This book was designed and typeset by David LeBlanc in Adobe Garamond 11.5/16 in Montreal, Quebec

To the Memory of
My Dear Sons

Andrew Alexander
† January 3, 1995

Timothy Andrew
† January 6, 1995

Always in my heart and mind

Contents

Documents

Illustrations

Acknowledgments

Although relatively short, this book has been long in the making – longer than I care to admit. I am thankful that the assistance, support, and encouragement of friends and colleagues alike sustained my interest over the years. To the following, in one way or another and in ways they may not even know, I owe a debt of gratitude: I.W. Bardyn, Dorothy Bittner, Adrian Boyko, Mary Carlton, Kyle Christensen, Kevin Corrigan, Wilfrid Denis, Jim Farney, Gladys Gibson, Bonnie Landago, L.Y. Luciuk, Craig Mahowsky, Peter Melnycky, Hans Michelmann, Myron Momryk, Desmond Morton, Kiran Mulla, David E. Smith, George T. Smith, Peter Smith, Don Story, Francis Swyripa, Myroslaw Tataryn, and Elaine Zerr.

Archival research, generally speaking, is expensive. Happily, I was fortunate to have enjoyed the patronage of several institutions and agencies, including the Yuri and Oksana Fedyna Fund at the Chair of Ukrainian Studies Foundation, the Canadian Institute of Ukrainian Studies at the University of Alberta, and St Thomas More College, University of

Saskatchewan. Their welcome support proved critical at various stages of this undertaking; to them all I wish to express my sincere appreciation by acknowledging their generous assistance.

The numerous period photographs in this book help to augment the text in important ways. Permission to use these images was kindly granted by the following organisations and individuals: the National Archives of Canada, the United States National Archives, the Provincial Archives of Alberta, the British Columbia Archives and Records Branch, the British Columbia Ministry of Transportation, the Glenbow Museum and Archives, the Whyte Museum of the Canadian Rockies, the Yellowhead Museum and Archives, Mr Walter Doskoch, and the late Professor J.E. Foster. I am much obliged to them. I also wish to extend a note of thanks to Mr Byron Moldofsky and the cartographic staff at the University of Toronto for preparing the finely crafted map that accompanies this book.

Several grants were received to offset the costs associated with the publication of this book. They include generous support from the Publications Fund and the Office of the President, St Thomas More College, the Chair of Ukrainian Studies Foundation, and the Publications Fund, University of Saskatchewan. The generosity of these agencies is deeply appreciated.

My relationship with the staff at McGill-Queen's University Press has been enormously satisfying. The interest of the editors in the subject, the kind words of encouragement, gentle prodding, and the careful shepherding of the manuscript through the final editorial phase were just the right ingredients needed to see the book through to print. For this I wish to express my heartfelt thanks to Philip Cercone, Joan McGilvray, Joanne

Pisano, and especially John Zucchi, all of whom afforded me great consideration while offering friendly advice along the way. Ron Curtis has served as my copy editor on a number of projects. In each case, as in this, his keen sense of judgment and stealthy use of the pen has improved the text greatly. His is unquestionably a talent much to be admired and to which I am clearly indebted.

Finally, it is with great humility and a profound sense of thanksgiving that I offer, however inadequately, my deepest and sincerest all to Danya whose faithful and cheerful self has been my Gibraltar and to my darling young son Christian who through his exuberance, dedication, and extraordinary creativity has set an example for his much humbled and admiring father.

Introduction

Canada's first national internment operation during the Great War is neither well known nor sufficiently understood, in part because it has not been the subject of much study or evaluation. This volume seeks to address this shortcoming by briefly exploring several thematic issues and questions that underscore the complexity of the experience, including the status of enemy aliens (and minorities more generally) during wartime; the role and responsibilities of government in moments of national crises; Canada's specific treatment of civilian prisoners of war during the Great War; and, finally, the nature of Canada and its ambitions at the start of a century that it claimed as its own.

The episode, characteristically speaking, was surreal, partly because of the ambiguous and at times conflicting nature of the experience. As the country moved along the road to nationhood, the enemy alien – by reason of birthplace – was consciously left outside the national project. That the enemy alien posed certain difficulties was to be expected, given the uncertainty and confusion brought on by war. But what is surprising is how local and largely incidental concerns and needs substituted for national policy. The Great War presented both a challenge and an opportunity for Canada, as a quasi-sovereign nation-state. In an exercise in national self-definition, the policies the leadership chose would either add to or detract from this process and goal. The argument here is that the national leadership responded to the enemy alien question, in a way that was weak, if not short-sighted, especially in light of the historical challenge.

This argument is perhaps best understood by looking at the official attitude toward the use of civilian internment labour on public works projects in the mountain hinterland of the Canadian West, a singularly important feature of the overall story. Although various belligerents used prisoner of war labour, especially in the latter stages of the global conflict, the *policy* in Canada of forcibly and systematically using civilian internees to work on public works projects during the war, a compensatory measure designed to appease various local interests while meeting government needs, was by all accounts exceptional. The compulsory use of civilian aliens of enemy nationality as prisoner of war labour would underscore the widespread belief that these were individuals who were outside the community and therefore not quite within the protection of the law. That their fate would ironically be bound up with other political objectives, the building of the national parks in particular, made it all the more remarkable and unusual. It is a meaningfully Canadian story, meriting, as it does here, special attention and further study.

And yet, it is also fair to say that despite some of the more unusual aspects of enemy alien internment in Canada during World War I, as a general proposition, it paralleled developments elsewhere. The internment of civilian prisoners of war was not uncommon, and the Canadian story, therefore, is as interesting for what it shares with as for how it differs from other experiences. Indeed, the internment of civilian enemy aliens as a recurring theme in the Great War heralds the mimetic irony that colours much of the twentieth century, in which romantic and heroic notions of conflict are at odds, sometimes absurdly so, with the pathos of the human condition in war.

This irony is nowhere better illustrated than in the bittersweet reflections on war in the twentieth century in which accounts of personal courage and self-sacrifice contrast with the cold, cruel, and often grim nature of mechanized warfare. Although it was not the first introduction to the modern requirements of war-making, the Great War would make the first full statement about the new potential of warfare. As an integral part of that experience, the internment of civilian enemy aliens would add to the modern vocabulary of war, hinting further at the more distressing aspects of a future total war. The Canadian experience contributes, if only in a small way, to this expanded appreciation of the meaning of war in the twentieth century.

As for the human condition in war and the circumstances that frame it, the story, perhaps, can best be told by those who were directly involved. For this reason, primary documents were included in the following chronicle, many of which are first-person accounts that not only communicate in highly accessible terms the layered and textured character of the experience but also speak eloquently to the human drama in the story. Sincere and often frank, these statements offer an honest portrayal of enemy alien internment in Canada, the authenticity of which is reinforced by period photographs whose rarity was assured by a military prohibition on taking pictures of the operation.

All the documents used in this volume were edited sparingly to preserve the flavour and character of the message contained within each. With minor exceptions (for example, where the original meaning may have been compromised), no attempt was made to correct for grammar or syntax. In addition, the names reproduced in the prisoners' rolls in the appendix, which have been culled from countless archival documents, appear in the original documents as variants. In the absence of a single definitive roster to use as a reference, one variant was chosen for each identifiable prisoner. Prisoners were assigned prison numbers, but some numbers could not be identified in the documents. They were therefore not included in the prisoners' rolls in this book.

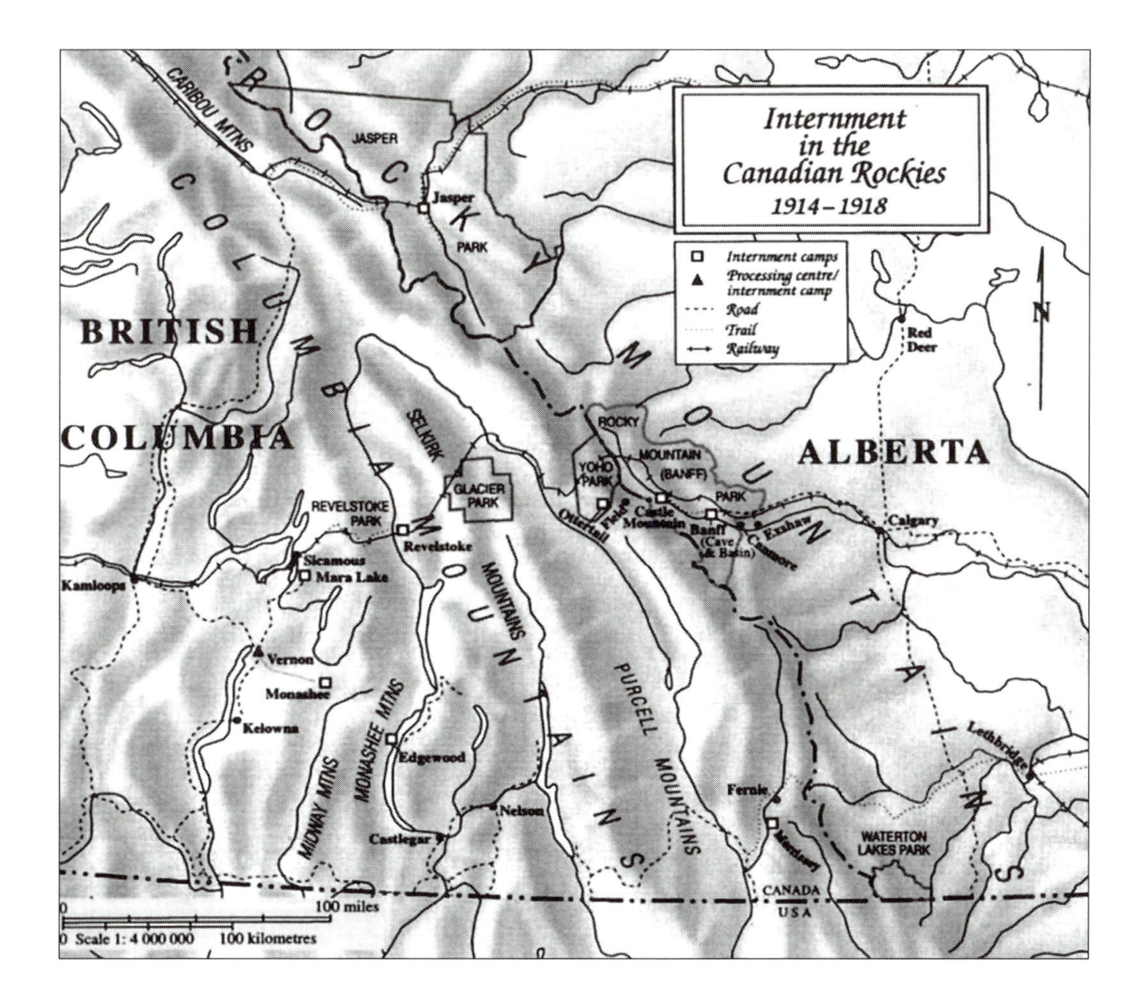

Internment in the Canadian Rockies 1914–1918
Internment camps
Processing centre/ internment camp
Road
Trail
Railway
N
BRITISH COLUMBIA
ALBERTA
CARIBOU MTNS
ROCKY MOUNTAINS
COLUMBIA MOUNTAINS
SELKIRK MOUNTAINS
PURCELL MOUNTAINS
MIDWAY MTNS
MONASHEE MTNS
JASPER PARK
Jasper
Red Deer
ROCKY MOUNTAIN (BANFF) PARK
YOHO PARK
GLACIER PARK
REVELSTOKE PARK
Revelstoke
Otter
Field
Castle Mountain
Banff (Cave & Basin)
Exshaw
Canmore
Calgary
Sicamous
Mara Lake
Kamloops
Vernon
Monashee
Kelowna
Edgewood
Nelson
Castlegar
Fernie
Morrissey
Lethbridge
WATERTON LAKES PARK
CANADA
USA
0 100 miles
0 Scale 1: 4 000 000 100 kilometres

Enemy Aliens, Prisoners of War

Guard of honour, govenor general's visit, Castle Mountain

1

Nation-Building and the Great War

IT IS AN ARTICLE OF FAITH that Canada came of age during the Great War. Indeed, such certainty is associated with this statement that it is now part of the historical canon. Yet before 1914, despite efforts at moulding a Canadian nation, success in that endeavour remained for many a distant, if not an elusive, goal, and for genuine sceptics it was, judging by the available evidence, quite simply impossible. Consequently, when the war provided an opportunity to demonstrate national resolve, Canada's political leadership, particularly in English-speaking Canada, embraced that opportunity as both a boon and a test of national will and purpose.

During the course of the war, for reasons both private and public, a great number of Canadians responded to the call to arms and through personal sacrifice gave credence to what Sir Robert Borden described in his tribute at the conclusion of hostilities as Canada's "proud and glorious history of participation." The European conflict was reassuring: it showed that there was a Canada worth fighting for and in whose name young men were willing to die. In many ways future generations of Canadians have come to accept and honour the sacrifices of the Great War as a testament to Canadian nationhood. Vimy Ridge, Ypres, and the Somme, for example, are part of the idiom of the national mythology, and Vimy Ridge, in particular, is now a pivotal event in Canadian history; at the battle of Vimy Ridge, it is often said, Canada came of age.[1]

There is no disputing the importance of such momentous events in the life and history of a people. Victories and heroes are essential in shaping historical identity. But the magnitude of these achievements, while giving them force and effect, also tends to obscure the complex nature of the nation-building process, detracting as it does from other routine, albeit no less important, aspects of the larger struggle for national affirmation. Canada's claim to political maturity should therefore not be judged solely in terms of the heroic deeds and valiant effort of young men and women in foreign lands. Rather, it must also be measured in the light of decisions that had as their specific aim and effect the building of a Canadian nation.

At its very core Canadian nation-building has been a conscious political act. From the outset various political decisions were intended to create a Canadian identity, strengthening the social, economic, and political bonds that would underpin a nation in the making. And yet with the decisions would come choices that were not entirely clear in their consequences at the time, but choices nevertheless. It is these choices that hold the greatest interest for us, since they illustrate the type of community being built and, perhaps more importantly, the costs that would and, it was felt, could be incurred as a result. Without detracting from the importance of the fight overseas for Canadian nationhood, a fuller appreciation of the complexity of the choices made at home provides a more complete understanding of Canada's attempt to define itself at the start of the twentieth century.

Canada's decision to enter the European conflict, although not an entirely independent one, met with little resistance in official Canadian political circles. The war, which was seen as a national test, would provide an opportunity for Canada to claim and secure its independent interests while establishing an identity of sorts in the process.[2] It has, of course, been difficult to completely distil Canada's role from that of Britain during the Great War, but fundamentally the image of Canada at war – epitomized by the Vimy Ridge metaphor – is an image of a nation struggling in common spirit toward a shared goal. But the real story is much more complex than this simplism. Indeed, as young Canadians were "going over the top" at Vimy Ridge, playing their role in Borden's "glorious history," another chapter of the Canadian national drama was being played out at home – the detention of thousands of immigrant arrivals from East Central Europe in Canada's first national internment operations.

Invited to people the Canadian West, they had been part of the great wave of settler-migrants at the end of the nineteenth and in the early twentieth century. But in August 1914, as un-naturalized residents and nationals of countries now at war with Canada, an estimated 120,000 were also designated as "enemy aliens." In what was considered to be a situation of peril and threat, and despite public assurances that persons respecting the law were entitled to its protection and had nothing to fear, 7,762 resident enemy aliens would be interned as prisoners of war.[3]

Public Notice to Enemy Aliens

It has come to the attention of the Government that many persons of German and Austro-Hungarian nationality who are residents of Canada are apprehensive for their safety at the present time. In particular the suggestion seems to be that they fear some action on the part of the Government which might deprive them of their freedom to hold property or to carry on with business. These apprehensions, if they exist, are quite unfounded.

The policy of the Government is embodied in a Proclamation published in the Canada Gazette on 15th August. In accordance with this Proclamation restrictive measures will be taken only in cases where officers, soldiers or reservists of the German Empire or of the Austro-Hungarian Monarchy attempt to leave Canada or where subjects of such nationalities engage or attempt to engage in espionage or acts of a hostile nature or to give information to or otherwise assist the king's enemies. Even where persons are arrested or detained on the grounds indicated they may be released on signing an undertaking to abstain from acts injurious to the Dominion or the Empire.

The Proclamation after stating that "there are many persons of German and Austro-Hungarian nationality quietly pursuing their avocations in various parts of Canada and that it is desirable that such persons be allowed to continue in such avocations without interruption," directs as follows:

> That all persons in Canada of German or Austro-Hungarian nationality, so long as they quietly pursue their ordinary avocations be allowed to continue to enjoy the protection of the law and be accorded the respect and consideration due to peaceful and law-abiding citizens; and that they not be arrested, detained or interfered with, unless there is reasonable ground to believe that they are engaged in espionage, or engaging or attempting to engage in acts of a hostile nature, or are giving or attempting to give information to the enemy, or unless they otherwise contravene any law, order in council or proclamation.

Thus all such persons so long as they respect the law are entitled to its protection and have nothing to fear.

Extra Canada Gazette, Ottawa
2 September 1914

Escorting prisoners, Victoria, British Columbia

At first glance these images seem to have little in common. But the decision that resulted in the internment of thousands of would-be Canadians was also in a real sense intended to address the problem of nation-building. Those who were interned – primarily resident aliens identified as enemies by reason of birth – were also in a metaphorical sense sacrificed, as the record would show, but in this case for the sake of public order and national solidarity, ingredients much needed in the making of the modern nation-state. The decision was a political calculation motivated by immediate needs but also made with the implicit understanding that in the larger balance of things some aspects of the nation-building process outweighed others. Little thought was given to the consequences of the policy of internment, however, even though it was known that the internees would one day have to be reintegrated into civil society.[4]

Without question internment reflected the augmented role of the state in the twentieth century, but as a political choice it also fell within the long tradition of nativist hostility and suspicion toward the migrant worker in Canada. It is, of course, one thing to speak of societal intolerance and quite another to identify the state as a source. But when, under the pretext of security, Canadian authorities used internment to deal with both the relatively benign problem of ethnic unemployment and the accompanying social tension, they were acting in accordance with the political and cultural ethos of the day – the state was identified with a homogeneous people, which implied that there were those who belonged and others who did not, or at least not until they had been fully assimilated and versed in the norms and attitudes of the majority. After all, as one political notable, J.S. Woodsworth, put it, these were "strangers within our gates." The ideal nation-state was an undifferentiated whole. If it did not exist, then it would be created.[5] Canada, in this sense, was still a nation in the making, and because the issue of birthright was magnified by fear of the conflict that lay ahead, rights were thought not to apply to those who were yet to be accepted into the body politic. As for the recently naturalized, their rights were applied only begrudgingly.[6]

Because they were perceived to be outside the community, enemy aliens would be subject to different standards of justice. Indeed, when civilian enemy aliens were interned as prisoners of war, they were, in effect, denied rights to which they were entitled as residents of Canada, well-established

"Why Should They Be Refused the Same Treatment?"

I am in receipt of a telegram from our Agent at Vancouver reading:

> Representations made by American Consul General Mansfield who is also the German and Austrian representative that at present some twenty Germans and Austrians absolutely destitute and numbers increasing daily. Associated charities refuse principally through lack of funds offer any assistance. Explains unable secure any information from representatives Austria or Germany as to what disposition or arrangements should be given such cases. Points out advisability some action be taken by Government first trying secure word from Germany or Austria through proper channels if any funds alleviate distress. In the alternative as many are becoming desperate will take drastic measures by robbery or otherwise even to the extent of becoming wilfully inmates of jails to prevent starvation. Is suggested such cases distress be placed in military prison similar to active military reservists. Have interviewed Colonel Duff Stuart here who primarily has requested matter be placed before you for consideration or what disposition you may deem advisable.

I do not suppose it will be possible to make through the usual channels any representations to the German or Austro-Hungarian Government and that there is little use in trying. If the only objection to these people is that they are destitute and unable to provide for themselves, I can scarcely see why they should be refused the same treatment as other immigrants would receive under similar circumstances. I take it that they have been neutral in their attitude, so far as returning or attempting to return to Europe is concerned, and that Mr Reid thinks they may take forcible measures only on the ground that they are suffering from a lack of food. I cannot see why they should be treated as military prisoners unless they have done something to bring them within the terms of Clause 2 of the Proclamation of the 15th ultimo.

W. Scott, Superintendent of Immigration, Ottawa, to J.A. Coté,
Deputy Minister, Department of Interior, Ottawa
23 September 1914

National Archives of Canada
Record Group 76, vol. 603, file 884866 (3)

"Local People get Suspicious"

Yours of August 18th to hand and noted. In reply would say that the copies of Undertaking Forms have not come to hand as yet but when they do I will attend to this matter as requested. Please send them at once.

Now Mr Scott I have a request to make of you not for the sake of myself but for the benefit of the country at large. At the village of Bruderheim, the Postmaster is a German and a bad one at that and in my mind he should not be in that position. I would kindly ask you to see that he is removed as soon as possible.

As soon as the Forms come to hand I will interview him and find out a few things regarding the orders that I have received from the department.

J.J. Libbey, Immigration Agent, Fort Saskatchewan, Alberta
to W.D. Scott, Superintendent of Immigration, Ottawa
4 September 1914

Referring to your recent letter in which you call attention to one G. Obertheur, a German postmaster at Bruderheim, who you think should be removed from his position, I may say that the Deputy Postmaster General, to whom the matter was referred, has made an investigation and I am informed that Obertheur has been a resident of Canada for 21 years and a naturalized citizen. He apparently has no sympathy with Germany and has not been in communication with anyone in that country for the last seven years.

I think it would be well if agents of the department would be very careful not to arouse any ill-feeling or disturbance in the case of Austrians or Germans residing in Canada.

Oftentimes, the local people get suspicious and cause a great deal of annoyance to one who, while born in one of these countries, has become a resident or citizen of Canada and may be as loyal as those within the British Empire. The very fact of such people being in Canada is to some extent an argument in their favour, since if they had been satisfied with their own country they would not have left it.

W.D. Scott, Superintendent of Immigration, Ottawa
to J. Libbey, Immigration Agent, Fort Saskatchewan, Alberta
2 October 1914

National Archives of Canada
Record Group 76, vol. 603, file 884866 (2 and 3)

rights that were recognized in international law and practice. Furthermore, once made vulnerable by their much-reduced legal standing, as prisoners of war they would be considered expendable by state authorities, who felt little compunction in deploying them wherever they were needed. They were used primarily on government works projects in the Rocky Mountains of the Canadian West – notably in road construction for the provincial government of British Columbia and, more particularly, in building up the infrastructure of various national parks.

The decision to use them in the development of Canada's Western national parks relates to the nation-building theme in a paradoxical way. From their inception in 1885, pressing commercial and economic interests encouraged the development of the parks. Significant also, however, was the idea of preserving the wilderness as a national resource and creating a symbol of national bounty and potential. This idea has, of course, now become a familiar leitmotif, strengthening the notion that the essence and meaning of Canada and Canadian identity is inextricably linked to the geography of the land. At the time, however, it was simply an idea. It would take committed parks officials in the Department of Interior who were armed with a sense of purpose and a certain single-mindedness to carry it through. When the war effort threatened to retard the work being done in the national parks, Dominion parks authorities would turn to internment labour for a solution.[7]

Significantly, no objection to this plan was raised. A pool of labour was available, and it would be used. Strictly monitored and enforced, internment labour then became of basic importance at this critical stage, providing some assurance that construction in the parks would proceed relatively unimpeded during the war. The park as a national symbol would continue to be developed at all costs. But ironically, the parks were developed without regard for that other political objective – which was very much part of a wider program and just as important – the building of a nation.[8]

Interned enemy alien labour could be used in Western Canada's Dominion parks and the British Columbia interior only because the status of the enemy alien internee had changed to that of a de facto military prisoner of war. The Hague convention, although silent on the treatment of enemy aliens, included a clause allowing, under certain conditions, the use of prisoner of war labour if what they worked on was unrelated to the war effort. Critically, the government of Canada was therefore allowed to use enemy alien labour if the individuals in question were in fact bona fide prisoners of war. That interned civilian enemy aliens could be something

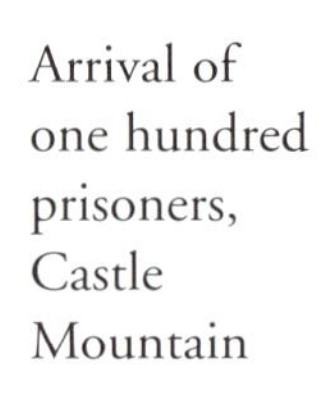

Arrival of one hundred prisoners, Castle Mountain

"They Will Be Treated as Soldier Prisoners of War"

With reference to your letter of the 8th instant enclosing a copy of telegraphic correspondence with the Governor General of the Commonwealth of Australia, I am commanded by the Army Council to acquaint you, for the information of Mr Harcourt [Colonial Secretary], that civilian enemy subjects who are detained, are provided with free accommodation, food and clothing on the same lines as soldier prisoners of war. They are not eligible for any pay, except working pay, and for this on the same conditions and at the same rates as soldier prisoners.

Civilians who are of a social position corresponding to that of officers may be allowed the same treatment regarding messing and accommodation as officer prisoners, provided that they wish for such preferential treatment and are willing and able to pay for their food and clothing out of their own pocket. This superior messing is paid for at a daily rate which in the United Kingdom is roughly from 4 to 5/- a day. If civilians of this class are unable or unwilling to pay the daily sum agreed, they will be treated, as messing and attendance, on the same lines as soldier prisoners of war. If possible, however, care is taken to segregate civilians of higher social standing and to give them a slightly superior accommodation, for which no extra payment is required.

B. Cubitt, War Office (London)
to L. Harcourt, Under-Secretary of State, Colonial Office
17 December 1914

National Archives of Canada
Manuscript Group 26, vol. 207, reel C-4441

other than military prisoners of war of course only complicated matters. So the question of their status as civilian internees and the responsibilities due to them was, for the most part, either ignored or denied.

Internees were forced to labour under trying conditions, even though the Hague Convention prohibited the crude exploitation of prisoners of war. The situation was truly extraordinary, in part because of the legal limbo in which the civilian enemy alien as prisoner of war found himself.

Under the war convention, military prisoners of war were accorded certain rights. Interned enemy aliens, however, with the exception of former reservists, were not military prisoners of war: they were noncombatants and as such, the rights guaranteed under international rules of engagement did not apply to them.

Generally speaking, the legal conundrum facing the enemy alien as POW was implicitly understood by the representatives of neutral countries. Tasked with monitoring the conditions under which prisoners of war laboured, they were inclined to view the problem as a domestic and not an international one. Consequently, since they were reluctant to press Canada to abide by international covenants that had very little to say about enemy aliens as so-called prisoners of war, they turned a blind eye, although confiding privately in various official memoranda their reservations about Canada's internment policy. In short, the ambiguous legal status of the internee – neither military prisoner nor civilian enemy alien – allowed an atmosphere of passivity to take hold, encouraging abuse and negligence and enabling the interned alien to be used as seen fit.

The view was widespread, including among the political authorities, that there was nothing unusual in internment labour. But this conclusion followed logically only if the interned were not considered part of the community. The paradox, of course, was that by virtue of their settlement they were already an indisputable part. Moreover, the evidence suggests that they were recognized as such. There was, for instance, no concerted effort to have all of them returned as a group to their countries of origin at the conclusion of the hostilities. Nor, with few exceptions, were the internees interested in repatriation. Their home was Canada.

New arrivals, Castle Mountain

To resolve the paradox, enemy aliens who were interned would have to be thought of as existing outside the community – at least for the duration of their internment. But if they were not part of the community, they were also, by extension, not entitled to its protection. This decision revealed much about the nature of the nation being built and the costs that, it was felt, could be incurred. The irony was that the internees would be forcibly put to work on a project that would assume great symbolic importance for the nation at a time when the leadership of the country was exhorting its people to struggle together in the pursuit of higher goals.

2

Between Ambition and Threat: Canada and the Problem of the Enemy Alien

WHEN CANADA FOUND ITSELF AT WAR with the Triple Alliance in August 1914 security measures were immediately implemented. Directed at those populations whose origins could be traced to the territories of hostile states warring with Canada, the initial orders in council, dated 7 and 13 August, respectively, set out the conditions that would determine the relationship between the state and enemy aliens: those who abided by certain imposed restrictions and pursued their ordinary avocations peaceably would neither be interfered with nor affected in any way. No one questioned the legitimacy of the cabinet orders, if only because they were well within the constitutional purview of the national government to ensure the security of the state, a practice widely recognized and sanctioned in international law. Given the political moment, the parameters seemed reasonable. And however uneasy the relationship with enemy aliens, it was at least defined, or so it appeared.

On the road, Banff

Within two weeks of the initial orders in council the federal minister of justice, C.J. Doherty, introduced a bill in parliament granting the federal cabinet extraordinary executive powers. The new War Measures Act, as it was called, granted the governor in council unparalleled authority to exercise sovereign powers, severely curtailing traditional rights, including *habeas corpus*. On the face of it, the focus of the War Measures Act was security, with various sections of the act proscribing the activities of unnaturalized aliens in a number of security-related areas. But perhaps the most striking feature of the orders issued under the statute were the broad interpretive latitude and discretionary powers awarded local agents acting on behalf of the state. Suspicion of illegal activity alone was sufficient cause for search, arrest, and detention. Indeed, although local authorities were instructed to make due enquiry to ascertain the facts and exercise judgment, the open-ended nature of the instruction ensured that a substantial degree of subjective opinion would be the determining factor in any particular case. What then is to be made of the act?

Its meaning is perhaps best understood by keeping in mind the remarks of the justice minister during the first reading of the bill before the House of Commons.[1] Charles J. Doherty, the minister, asked the House to give the government "the power to exercise its own judgment as regards the requirements that the various conditions which we have to apprehend may call upon us to meet." In arguing for unrestricted power, Doherty attempted, moreover, to persuade his listeners that "It was for every man to do that which lies within his power, on behalf of our country whose fate … is hanging in the balance." The minister of justice voiced a sentiment that was widespread, delivering words that implicitly communicated, given the emergency nature of the domestic legislation, the existence of an internal threat. The enemy alien designation, a classification reserved for un-naturalized immigrants, clearly identified a group that, being characteristically outside the wider community, satisfied the notion that a domestic threat in fact was real. As for those inside the community, they were being exhorted to stand in its defence. The minister's remarks would strike a resonant chord with the House, the bill passing expeditiously with virtually no debate.

"He Might Give Us Trouble"

I have the honour to state that I had previously interned this man [Thomas Koch], but, on interviewing him yesterday, he informed me that he has bought a quarter section of land near Lampman at $20 an acre. I wired Constable Robson as per the copy of the telegram forwarded to you. Constable Robson replied:

> Thomas Koch has quarter section land and four horses on section eighteen, four, four, west second. Stop. Mrs Glasser is looking after same. Stop.

From the report, it is clear that when this man was arrested he was acting in an ugly and suspicious manner. Do you think it advisable to send him down to Brandon? I think he should go, since his farm is being looked after. I do not see that he would be put to any trouble, while he might give us trouble.

RNWMP Superintendent, Regina District, Saskatchewan
re: Thomas Koch, interned at Morrissey Camp
27 April 1915

National Archives of Canada
Record Group 18, vol. 1770, file 170

In the Canada of August 1914, un-naturalized aliens did not collectively pose a direct threat to the security of the state, at least not in the conventional sense of the term. There was, of course, widely shared concern about alien reservists, who had to be prevented from joining the fight overseas, as well as about the possibility of saboteurs and fifth-column activity, all of which prompted the quick introduction of appropriate security legislation and necessary security checks at the Canada-U.S. border. There were also impolitic statements voiced by naive men – individuals whose conservatism and narrow worldview conflicted with the demands of the moment – and the petty jealousies of ambitious community gadabouts whose activities proved less than reassuring.[2] But this was the stuff of scandal, not treason. Indeed, the ill-considered remarks and the parochial politics of opinionated editorialists, publicists, and community leaders alike were not the concern of officials, since there was little in the way of public sanction or legal consequence for the statements that were made.[3] Rather, what did concern officers of the Crown more immediately were those aliens designated by law as the enemy whose foreign appearance, demeanour, and sullen look were grounds alone for suspicion and arrest.[4] The threat they posed, however, was more

Internees at Sulphur Springs, Banff

abstract than real. Their "threat" was to the historical enterprise of nation-building and, more precisely, at this particularly crucial moment in history, to the commitment of a nation being called upon to participate in the struggle ahead.[5]

Indeed, when the justice minister spoke of the country whose fate was hanging in the balance, it was not the fear of imminent invasion that animated the passion behind his words. Nor was it the prospect of losing the war on the field of battle. But for C.J. Doherty the stakes were just as high, since he was speaking of no less than the future of the country in the balance of history. What historical judgment would be passed on Canada in the great struggle ahead? Would the country rise to the occasion and contribute to the war in a way that would be worthy of recognition and independence? In the shadow of Empire, these were the grand questions that would set Canada's course for the twentieth century. Against this, however, was the improbable contrast of the resident enemy alien, who offered little in the way of inspiration. Sullen and lowly, without work and the necessary papers in hand, as he stood before a registrar or magistrate, he was not only a social and economic burden but a liability, a problem for which political solutions would have to be devised.[6] Once the un-naturalized alien had been designated a domestic enemy and once extraordinary powers had been delegated to local authorities, it was inevitable that the foreigner as enemy alien would become the target of authorities who acted out their worst fears and prejudices, all the while believing in the just nature of their cause.

For societies built on immigration, nativism is an insidious phenomenon often exacerbated by stressful economic, social, and political conditions. Those conditions came together in 1914 in an unfortunate series of events and circumstances that worked against normal, let alone harmonious, relations between peoples. The ethnic underclass was economically pressed during the recession of 1913–14, but the war, with its emphasis on corporatism, as well as productive and socially useful labour, severely tested the limits of domestic acceptance of the ethnic worker. During the war it was in fact a personal catastrophe to be both ethnic and unemployed: the unemployed alien was frequently, though somewhat mistakenly, perceived as a drag on and a liability for the war effort.[7] From this way of thinking it was but a short step to equating the activity of the unemployed foreigner, whose status suggested questionable allegiance, with the potential for sabotage. The war would therefore prove disastrous for the many thousands of recent East-Central European immigrants who were unemployed and had no means of support. As for the guarantee in the order in council of 13 August offering protection to enemy aliens who pursued their ordinary avocations peaceably, it was in fact a meaningless gesture for the unemployed enemy alien.

State policy, however, was not simply following the vagaries of domestic public opinion. After all, even if state officials were not fully able to deal with an overtly hostile public, the political leadership was still in a position of authority. Faced with the clamorous demands of xenophobes who urged a general detention, the American president, Woodrow Wilson, by comparison, introduced legislation that was specific, limited, and focused, resulting in a program of internment that was selective and restrained.[8] In this sense, the political authority invested in Canadian officials was neither responsibly nor fully exercised, since their policy choices

An Official Perspective

The Local Committee made a full statement of the case at a meeting presided over by the Hon. E.G. Prior [Member of Parliament, Victoria]. The Committee demanded the internment of all "alien enemies," and a government investigation into charges against naturalized citizens of German birth. After a discussion lasting nearly three hours, in which every phase of the question was ventilated, Colonel Macpherson [Internment Directorate, Ottawa] stated that, personally, he was in full sympathy with all the requests made. For his own part, he would like to see every "alien enemy" interned. He would also like to see a more drastic treatment of naturalized citizens of German birth who had made themselves obnoxious. However, with respect to the former proposition, he said that the government policy was not one of general internment because of the vast numbers which would have to be dealt with. He mentioned the fact that there were nearly a quarter of a million in Alberta and Saskatchewan and it would be physically impossible to intern such numbers. The policy of the government was to intern only those who had given occasion for offence, and against whom some specific charge could be proved.

The Committee assured Colonel Macpherson that this attitude was not at all satisfactory and would not meet the wishes of the people of Victoria. They regarded every "alien enemy" as a potential source of danger, and as wholesale internments had been found necessary in other parts of the Empire there was no reason why they should not take place here, the case of Alberta and Saskatchewan having nothing to do with British Columbia. Victoria was not asking for a national policy on this question, but for a local policy to meet conditions which were unlike any other city in Canada. In no other city had there been such an outburst of public feeling with destruction of property and danger to life.

Colonel Macpherson could not give any assurance that this view would be adopted by the government, but promised to report the wishes of the Committee. With reference to naturalized citizens of German birth, Colonel Macpherson was very emphatic in stating that nothing could be done except through the ordinary channels of Common Law. Canada was not under military law, and the only procedure was to lay a complaint with the police authorities in the ordinary way and to take the matter into court.

He admitted that some naturalized citizens of German birth were far more dangerous than "alien enemies," and all the more so because of the greater opportunities and the fact that they could screen themselves behind their naturalization papers. He held out no hope of special government action in such cases, nor did he promise to recommend a local investigation. One point which had long been in doubt was cleared up by Colonel Macpherson. He stated emphatically that Major Ridgway Wilson, the Internment Officer for British Columbia, had full authority to intern within the provisions of the Act without reference to Ottawa, and wound up by saying that if any members of the Committee or any citizen would lay a specific complaint against an "alien enemy," and prove it to the satisfaction of Major Wilson, then that officer would intern them.

Fernie Free Press
Fernie, British Columbia
June 18, 1915

Cause of Arrest

ILIA PETRASCHUK, No. 227 Banff. Age 21 yrs; Ukrainian; labourer. Captured at Calgary 6th of December 1915. Interned at Banff, 7th of December 1915. Is not a Reservist. Cause of Arrest: Not reporting for six months. Destitute. Received from Calgary Detention Barracks.

WASYL HUCULIAK, No. 228 Banff. Age 20 yrs; Ukrainian; labourer. Captured at Calgary 6th of December 1915. Interned at Banff, 7th of December 1915. Is not a Reservist. Cause of Arrest: Not registering and lying, saying he had registered when he had not done so. Received from Calgary Detention Barracks.

NYKOLAJ KOZMA , No. 233 Banff. Age 42 yrs; Ukrainian; labourer. Captured at Calgary 7th of December 1915. Interned at Banff, 9th of December 1915. Is not a Reservist. Destitute. Cause of Arrest: Not reporting and lying. Received from Calgary Detention Barracks.

NYKOLAJ WYNNYCZUK, No. 234 Banff. Age 45 yrs; Ukrainian; labourer. Captured at Calgary 7th of December 1915. Interned at Banff, 9th of December 1915. Is a Reservist. Cause of Arrest: Not reporting and lying. Destitute. Received from Calgary Detention Barracks.

PETRO JAWORENKO, No. 235 Banff. Age 24 yrs; Ukrainian; labourer. Captured at Calgary 7th of December 1915. Interned at Banff, 9th of December 1915. Is not a Reservist. Cause of Arrest: Not reporting and lying. Destitute. Received from Calgary Detention Barracks.

Prisoner Arrest Records Submitted for Release Consideration
31 May 1916

National Archives of Canada
Record Group 117, vol. 14,
file "Correspondence – Release of Prisoners"

were apparently more in keeping with the prevailing public mood. Instead of dealing practically and judiciously with genuine and clearly defined security issues, they cast the beleaguered alien in the role of an enemy and a threat. How is this policy to be explained?

Order in Council (PC) 2721, which was passed by the federal cabinet on 28 October 1914 under the authority of the War Measures Act, put in motion the process of civilian internment. The order provided for the creation of registration centres tasked with the security function of monitoring the personal status of the enemy alien and with the internment of enemy aliens who violated the various directives regulating conduct under this and other orders issued under the declaratory power of the government. These orders included the regular reporting before magistrates or local police of all enemy aliens who lived within a twenty-mile radius of designated registration centres (later amended to cover all enemy aliens over the age of sixteen, regardless of location); restricting movement of enemy aliens around security areas, notably railways and bridges; and preventing international travel of enemy aliens, specifically those who would leave Canada without an official exeat for the still-neutral United States. Although there was an important security dimension to PC 2721, the objective of this and several subsequent orders was the social and political control of an identifiable population. The immediate cause was the heightened concern over the unemployment situation among this group.

The economic recession of 1913 had deepened by 1914. In the spring of 1914 government economic forecasts projected an unemployment crisis of unprecedented proportions, affecting, in particular, the ethnic labourer as the most vulnerable part of the labour force. Protests and

Coming out for the roll call, Castle Mountain

labour unrest in the larger Western Canadian cities, accompanied by reports of alien workers travelling in groups between cities in search of work, were of concern in Ottawa.[9] The anxiety was further compounded when, with the outbreak of war, hundreds of enemy aliens, many penurious with no prospects of employment, headed for the international boundary but were prevented from crossing.

In late October 1914 Prime Minister Borden alerted London that the situation in Canada was becoming desperate, since it was predicted that from fifty to one hundred thousand individuals would be unemployed

"I Have No Evidence"

JOHN KORSHIWSKI

On cross-examination, under Section 2 of the Proclamation, I could learn nothing further, only that he was attempting to leave Canada for the United States. This man was working as a labourer for the "World at Home Circus," which was engaged at Brandon, Manitoba four weeks ago, and was travelling with the circus to Northgate on their way through to the United States. I arrested him on the arrival of the circus train at the GTP Railway depot at the village of Northgate, Saskatchewan on the international boundary. He had no papers or letters of any description, and was without money or personal property. He speaks fairly good English and says he has not been paroled. I have no evidence to prove whether his replies to the above are true or not. The men working on the circus corroborated his statement as to when and where he joined the circus.

Summary Remarks, RNWMP Crime Report, Weyburn, Saskatchewan
Re: John Korshiwski, interned at Jasper Camp
22 August 1915

National Archives of Canada
Record Group 18, vol. 1779, file 170

during the coming winter.[10] Although the idea of recruiting this large element into the armed forces in order to alleviate the problem was discussed, it was rejected. Borden's proposal to relax the conditions for out-migration, which would have allowed those wishing to leave for the United States to do so, was also rejected, even though there was little probability the aliens would ever reach their home territories and join the fight against the Allies. British authorities considered it an unwise precedent to allow them to leave and proposed detaining them.[11] Consequently, facing an unpredictable labour situation and the prospect of a

European war – but also constrained by Canada's quasi-colonial status – Canadian authorities felt compelled to act in a way that, within this context, did not threaten their basic assumptions about nation building.

Significantly, the Canadian prime minister's original proposal to allow aliens to leave the country was based on the consideration that the aliens had been "invited" to Canada. In the initial invitation there was promise and expectation, so why, suddenly, was there now a lack of empathy? Indeed, the economic problem posed by the enemy alien belies the similar and no less difficult predicament of the countless thousands of other naturalized and un-naturalized immigrants, "Britishers" and "non-Britishers" alike, who were now unemployed or who were faced with the prospect of unemployment.[12] That the enemy alien could be targeted without due concern, let alone sympathy – even though there was a moral obligation for having attracted and invited these people to Canada – had as much, if not more, to do with the national ambitions of Canada and the leadership that guided the ship of state than simple political uncertainty brought on by economic hardship.

This is not to say that there was not some cause for concern when thousands of foreigners gathered in Winnipeg in May 1914 demanding "work and bread," or a year later when, after suffering a horrible winter of unemployment, a column of unemployed workers ignored official threats and marched to the United States border in search of work, claiming it was "better to die of bullets than hunger."[13] But the concern was of a very specific type.

Although the Canadianization of the foreigner had always been thought of in official circles as an attainable goal in the long run, the unruly conduct and the impossible demands made by the aliens were unhelpful and unacceptable given the political moment. Although diligent policing could well have dealt handily with any disturbance, the real problem for officials was the alien character of this group, whose behaviour and attitude placed them at odds with the national agenda and distracted others from the purpose at hand. The disgruntled alien did not provide the stuff with which nations should enter into battle nor upon which the future would be built. This view was widely held, and it bolstered the unfolding state policy.

The conundrum of having once invited individuals who were now unwelcome was not lost on the national leadership. By casting the alien in the role of enemy, they effectively resolved this dilemma. Constrained by a colonial legacy and driven by national ambitions, government officials made a conscious decision, in their eagerness to move forward along the road to nationhood, to leave part of Canadian society behind. That it did not appear to weigh on their conscience was not surprising. There was little sympathy for an enemy with whom the state in a figurative, if not a literal, sense was at war.

Tallying the prisoners, Castle Mountain

3

Enemy Alien Internment: Obligations and Responsibilities

DESPITE THE CONCILIATORY NATURE of the order in council of 7 August, which guaranteed protection to those who abided by the law, hundreds of enemy aliens were arrested in the first weeks of the war, especially after a proclamation of 15 August had extended wide powers of arrest to the Dominion and Royal North West Mounted Police. They were at first detained in municipal lock-ups and provincial gaols, but as their numbers increased, local authorities secured additional public space to serve as detention centres, including the Immigration Building in Montreal (opened 13 August), Fort Henry in Kingston (18 August), the Fort Garry and Osborne Barracks in Winnipeg (1 September), Halifax's Citadel (8 September), the provincial government buildings in Vernon (18 September) and Nanaimo (20 September), Brandon's Immigration Hall (22 September), and the Exhibition Building in Lethbridge (30 September). All these facilities were in various stages of disrepair, pointing to the general state of unpreparedness. And because more prisoners were expect-

First arrivals, Otter

ed, it was clear that the ad hoc approach to detention would not suffice and that a more systematic tack would have to be taken.

On 28 October, therefore, C.J. Doherty, the federal minister of justice, presented his cabinet colleagues with a report outlining a comprehensive plan that recommended the registration and internment of enemy aliens. The policy called for all un-naturalized immigrants originating from the territories of states that were at war with Canada to report to local officials and carry state-issued papers on their person at all times. Individuals who did not register or who failed to appear regularly before a commissioned registrar and answer questions truthfully would be subject to detention, and if they violated the various government orders, they would also be interned. Furthermore, the right to trial was suspended. Significantly, the report not only called for the military authorities to make the necessary provisions for the maintenance of enemy aliens interned as prisoners of war, but it "would require such prisoners to do and perform such work as prescribed by them." The policy proposal was accepted and adopted as order in council (PC) 2721 – on the same day the minister had presented it to cabinet for approval.

PC 2721 was soon followed by order in council 2817, issued 6 November, which approved the appointment of a director for internment operations, the retired major general Sir William Dillon Otter.[1] Order 2817, however, was something of a formality. Otter had actually been summoned to Ottawa on 30 October, had been offered the position of director on the 31 October and had assumed his duties on 1 November. There was good reason for the expeditious manner in which the commission was approved. The earlier order in council of 28 October had provided local officials with the authority to get on with the business at hand, and by the end of November 1914 approximately 8,200 apprehended enemy aliens had been arrested and were being readied for internment.

Prisoners erect wire fence, Otter

Because of the growing volume of prisoners being processed more holding centres were required. They were obtained chiefly by converting several military installations into so-called receiving stations, including Toronto's Stanley Barracks (14 December), as well as the armouries at Niagara Falls (15 December), Beauport, Quebec (28 December), and Sault Ste Marie (13 January). The armouries at the last three locations were particularly needed, since scores of enemy aliens were being detained at these ports of exit in their attempts to leave the country. But larger, more permanent facilities were also required. The militia camp at Petawawa was made available for this purpose, while land grants were obtained from the Ontario and Quebec governments near Kapuskasing and Spirit Lake, Quebec, with the idea that camps could be built on these sites in short order. Petawawa began to operate on 10 December, while the first prisoner contingents were sent to the northern reaches of the Canadian hinterland of Kapuskasing and Spirit Lake on 14 December and 13 January respectively.[2]

Despite the efforts at securing the necessary facilities, not all of the 8,200 individuals who were detained could be interned during this transitional period. Consequently, when they had sworn the obligatory oath of allegiance, most were paroled. This large initial number was nevertheless significant, because it demonstrated not only the verve with which officials pursued their work but also the absence of real political or legal safeguards. The War Measures Act ensured this result, as did various specific government orders issued under the act. Clause 7 of the order in council of 28 October (PC 2721), which established the framework for registration and internment, for example, declared that "any alien of enemy nationality who in the judgment of the registrar cannot consistently with the public safety be allowed at large shall be interned as a prisoner of war." This clause, in particular, gave registrars relatively unregulated authority to interpret and decide on the security needs of the state. It would also provide an opportunity for unscrupulous types to callously profit from the misfortune of others, a few demanding bribes in exchange for dispensation. In the end the clause became a crucial provision affecting the fate of the many thousands of designated enemy aliens.

"My Belief"

IWAN MILAN

I have made this man a prisoner of war, charging him under Sec. 2 Sub-section (c) of the Proclamation of August 15, 1915 – "With being a subject of the Austrian monarchy, and engaged in attempted acts of hostile nature against us." The reasonable grounds for my belief being as follows: associating with men of his own country who are drilling like soldiers; wandering around to different points where he can gather information; associating with persons who are very bitter against us; has a bad character in that he would be one of the first men to do any act which would hurt either the military or the police, if by doing so he could earn a dollar of any kind; having no fixed abode; his movements are open to suspicion; and he has not a cent in the world, only the clothes he stands up in practically. In short, he has not done any work since October last, has no visible means of support, and was taken from a place of bad repute – re: agitators.

Summary Remarks, RNWMP Crime Report, Yorkton, Saskatchewan
Re: Iwan Milan, interned at Jasper Camp
15 January 1915

National Archives of Canada
Record Group 18, vol. 1779, file 170
Of Bribes, Law and Responsibility

Of Bribes, Law, and Responsibility

I understand that my brother's release was refused by General from Ottawa. I cannot see any reason why it was refused. I wish you would inform me why we are kept in here. Of course I know the charge because I crossed the International Boundary twice, but I did not know that it was against the law. Therefore, I should think that I am not guilty of the charge. I was out of work and money at that time. My brother put up $300 as security so that I may not be charged with anything. Police who arrested us said, at the time, that he was mad and that was why he arrested us. Should you want anything more in connection with this matter please call me to the office.

Prisoner John Marchuk
to Major Stuart, Commanding Officer, Castle Internment Camp
22 October 1916

Referring to your letter of the 31st ultimo, I have the honour to inform you that No. 589, P. Marchuk, was interned for crossing and re-crossing the International Boundary Line between Canada and the United States in the vicinity of North Portal, Saskatchewan. No. 590, J. Marchuk, was interned for the same reason.

This matter has been a subject of correspondence between Major D.H. Coleman, formerly the Officer Commanding, Internment Operations at Brandon, Manitoba as a result of ex-constable Waston being charged with accepting bribes for which he has already been convicted and sentenced to a term of one year's imprisonment.

RNWMP Commissioner's Office, Regina, Saskatchewan
to Major General Wm Otter, Director, Internment Operations
7 November 1916

National Archives of Canada
Record Group 6 HI, vol. 759, file 3565

The wide discretionary power given to registrars had obvious consequences. Although the majority of the initial 8,200 individuals who had been detained were paroled, with facilities and a routine now in place it was only a matter of time before the number of detainees increased once again. On 10 February 1915, those arrested and subsequently interned totalled 1,904. The figure grew to 2,752 at the end of March and doubled again to 5,088 by June 1915. In addition, 48,000 were reporting to local authorities. By 1 May 1916, 6,061 internees were confined behind Canadian barbed wire. As for those who had registered and were required to appear before a local authority – so that their activity could be monitored and regulated – some 80,000 had done so by this time.

Among the internees were merchant marines, primarily of German origin, whose uncertain status had led them to be transferred to Canada from the Caribbean at the request of British imperial authorities for the duration of the conflict.[3] There were also reservists, former nationals of the belligerent states, who, because of prior military experience, were eligible for foreign military service; international law identified this category of individuals as liable to internment. But among the final tally of 8,579 who were interned in Canada during the course of the war, the vast majority were un-naturalized resident immigrants who had arrived at the behest and invitation of the Canadian government to homestead in the prairie West at the start of the century.[4] Labelled "Austrians," they were primarily peasants hailing from Ukrainian ethnographic territories within the ramshackle Austro-Hungarian Empire, which was now at war with Canada. Homesteading had proved difficult, and a number had gravitated to the urban centres of the West in search of work and relief. As individuals existing outside the body politic, the lack of employment opportunities during the deepening recession in 1914 made them extremely vulnerable. They consequently became the primary target for arrest and detention and constituted the bulk of those who were finally interned.

Not surprisingly, in Western Canada, where many of these individuals were located, the number of arrests increased as municipalities faced with the problem of growing indigence looked to internment as a solution. For General Otter, who had a more conventional view of internment and its objectives, the willingness and ease with which municipal councils were able to relinquish their political responsibility was cause for dismay. He was completely confounded by the situation at the Lakehead, where in January 1915 local authorities successfully organized the arrest of eight

Prisoners' compound, Castle Mountain

Major General Sir William Otter on site, Revelstoke

hundred enemy aliens for the purpose of clearing both the bush in the surrounding township and the alien poor from the streets of Port Arthur and Fort William.[5]

What Otter failed to appreciate, however, was that the municipal governments had taken their cue from federal authorities, who had not only accepted internment as a legitimate policy option for dealing with the social effects of the economic crisis but, as in the case of the deputy minister of justice, E.L. Newcombe, had also described it without apology as a generous policy. In response to criticism Newcombe argued that by turning hungry indigents who had been abandoned by their own governments into prisoners of war and interning them, the government of Canada was able to provide for their keep. As for their rights, he was explicit: "it is according with our domestic system to employ, at such labour as they are qualified to perform, persons whether native or foreign who are cast upon the charity of the State, and [the minister] apprehends that neither the state of war nor any rule sanctioned by international convention or practice requires that destitute people of any nationality, when seeking relief from the State, should be immune from a similar requirement."[6]

Newcombe's argument was echoed in clause 10 of PC 2721, which declared that prisoners of war were to perform work required of them. It also resonated in the public's unquestioned acceptance of internment as a policy whose purpose was social relief; for example, Winnipeg's *Free Press* reported after PC 2721 was promulgated that the order authorizing internment was intended to relieve the distress of enemy aliens who were without work or who could not support themselves.[7] The argument was that the policy, by its own merits, was innocuous and that it was to the benefit of all parties concerned.

The same reasoning informed order in council 1501, which passed on 26 June 1915. The enemy alien, viewed as a competitor and a threat to native-born workers in a scarce job market, evoked considerable resentment and animus from an increasingly anxious and hostile public. Assessing the potential for social unrest, the minister of justice recommended through PC 1501 "the apprehension and internment of aliens of enemy

"What Will Be Done?"

Where are we at? Who knows? This about sums up the alien internment situation at Fernie … The provincial police interned the aliens on instructions from their superior – the Provincial Chief at Victoria acting on orders from the Attorney General's Department. The Province took the responsibility for locking them up and guarding them. Along comes the representative of the Dominion for internment matters, Colonel Macpherson, and refuses to accept the charges of the Province. He can see no legal authority for interning these men and so advises his headquarters. The Dominion Government says they are unlawfully detained. "Hold them!" says Attorney General Bowser.

Mr McNiven of the Labour Department comes here, looks over the situation and departs. Immigration Inspector Reid comes in to see what is what, hears that the Dominion absolves itself of responsibility, and makes a hasty get-away. A representative of the Federation of Labour comes from Winnipeg to see what has been done. The local lawyers refuse to act for the internees so they get outside counsel and habeas corpus proceedings are commenced. Arguments will be heard on the coast Monday.

According to reports in yesterday's Lethbridge paper, Colonel Macpherson suggested that the local men might be taken to that point for internment. This would indicate that he feels that the Dominion will ultimately assume care for these men. Attorney General Bowser will try to get the Dominion authorities to act consistently, and to do with these men as was done with those who were ordered interned at Nanaimo. Messrs Barnard and Green, MPs, have gone to Ottawa, no doubt in connection with these matters.

What will be done? Hillcrest has followed the lead of Fernie and the Britishers there have quit work pending the dismissal of aliens from the mines or their removal. As mentioned above, it's anyone's guess as to the outcome. Here is ours based on considerable confidence in the wisdom of the powers that be. The Dominion Government will realize that an extraordinary situation exists here as well as in the other mining camps, and will frame a new policy covering the situation. In keeping with that policy, the men at present interned here will be taken to a permanent internment camp in due course. The calmness and order which has characterized all the proceedings in this connection to date will be maintained on all sides and the incident will be quietly closed.

Fernie Free Press
Fernie, British Columbia
June 18, 1915

nationality who may be found unemployed or seeking employment or competing for employment in any community." Internment was used in this instance to bring about labour peace, and the policy was justified by arguing that the internees were protected in this manner. The stated rationale for order 1501 belied its real effect, which was to provide another legal avenue for arrest and detention. When miners at the Crow's Nest Pass Coal Company of Fernie, British Columbia, agitated for the internment of enemy alien co-workers, it was order 1501 that was invoked.[8]

The argument that enemy aliens were being interned for their own good was disingenuous. Unwilling to deal directly with nativist pique, the government issued various orders to alleviate local fears and appease the prejudices of the host society. The Canadian state had, in effect, relinquished its responsibility. Perhaps more critically, however, the policy of internment was initiated under a legal guise, when, in point of fact, the actions of the government of Canada were incompatible with the requirements of international law and practice. Indeed, when the deputy minister of justice, E. Newcombe, claimed that transforming enemy aliens into prisoners of war and subsequently interning them was both right and good, he was contravening some well-established notions of justice and law. Specifically, even if they were designated as *enemy* aliens, they still enjoyed certain rights under recognized international and national standards and practices.

There has been a long tradition of international agreement governing the conduct of war on land. Although within that tradition and before the summer of 1914 very little was said that directly addressed the problem of the status of the enemy alien in time of war, two guiding historical principles nevertheless applied. The first was a general principle that domiciled aliens were subject to the laws of the land under the same conditions as natural-born subjects. While reaffirming the alien's legal obligations to the host state, this principle also implied that an alien's substantive and procedural rights were the same as those of ordinary citizens of the country, less the special civil privileges and political rights reserved for citizens. Described as a national standard of justice, its transgression would provide the necessary justification in international law for the aliens' country of origin to intervene on their behalf.[9] The second principle was the specific historical right of free passage, or exeat, in times of crisis. This right,

Prison barracks, Morrissey

Rights and Privileges

No prisoner of war should under the present conditions accept any employment in Canada or in any British owned colony. Reason: In time of war no British subject is bound to pay any alien enemy for services rendered, or, if transacting business, neither bound to pay for the goods delivered during the war. Furthermore, no prisoner of war is bound nor can he be forced to pay taxes or duties of any kind. Taxes and interests have to wait for their continuation until after the war. Private property is not confiscable. Everyone interned has the right to claim damage if it was the fault of the government to keep them back at the outbreak of hostilities. The sovereign declaring war can neither detain the person nor the property of those subjects of the enemy who are within his dominion at the time of the declaration. They came into this country under the public faith. By permitting them to enter and reside in his territories, he tacitly promised them full liberty and security for their return. He is, therefore, bound to allow them a reasonable time for withdrawing with their effects, and if they stay beyond the term, he has the right to treat them as enemies. But if they are detained by an insurmountable impediment, as by sickness, he must necessarily, and for the same reasons, grant them sufficient time.

Extract of Correspondence, J.C. Woodward, American Consul
to unidentified prisoner, Mara Lake Internment Camp
27 August 1916

National Archives of Canada
Record Group 6 HI, vol. 752, file 3033

which was established through customary public law as part of the accepted tradition involving diplomatic protection of nationals abroad, allowed enemy aliens who wished to abandon their residence to do so within an allotted period of time.[10]

When Canada went to war in the summer of 1914, its actions respecting enemy aliens should have been governed by its adherence to these two principles. To be sure, it was difficult to adhere to them when entire nations were being mobilized for the conflict that lay ahead. It is often argued that liberal attitudes to the rights of enemies are entertained only at a nation's peril, and in this sense the War Measures Act reflected the new political reality of nations at war. Nevertheless the implications were also clear. Without political safeguards, the state's prerogative to exercise its authority would make for a troubling situation if rights were held in abeyance and much was excused in the name of national security.[11]

Despite the new demands of modern warfare, Canada's international obligations were clear; they were spelled out by various international practices and agreements. Canada was not alone here. Other states were similarly bound and attempted to work within the same parameters and to meet the same rudimentary standards established through historical practice. Although the record was mixed, especially among the front-line European states, they nevertheless made an effort at compliance, demonstrating an acute awareness among all the belligerents of their international responsibilities.[12]

In the end, what particularly distinguished the Canadian experience was the government's disregard of its fiduciary obligation to the newcomers, an obligation that flowed from the special circumstances surrounding the relationship that existed between the government of Canada and the recent arrivals. Unlike its European counterparts, the immigration policy of the government of Canada at the turn of the century actively recruited potential homesteaders for the prairie West. Once having invited them to Canada in a campaign promising opportunity, the government had, arguably, a moral and political responsibility to extend to these immigrants the same national standards of justice that were enjoyed by its citizens. If it was unable to provide that minimum standard, the government was obliged at least to recognize the right of exeat, to allow those who wished to leave to do so.[13]

Of basic importance, however, was the notion that immigrants could expect justice and that standards would be adhered to. Perhaps for this reason, those who were eventually interned found their transformation from enemy alien to actual prisoner of war so incomprehensible. Among the internees there was confusion and disbelief. "Prisoner of war" implied active participation in the conflict. For most, the European war was remote and had little to do with immigrants whose former minority status in a polyglot empire translated into political ambivalence at best.[14] Moreover, they had come to homestead not to engage in politics, and homesteading implied a commitment that they, in turn, could reasonably expect from the country they now called home. One individual who had been personally affected by internment operations would exclaim that it was hard for him to believe that Canada would intern its "own people." The anger behind those two pointed, yet well-chosen, words was fuelled by a heartfelt and profound sense of betrayal.

The war had the unfortunate effect of creating enemies of a certain segment of the population. Under the statutory powers of the War Measures Act, Canadian authorities had the legal right to intern enemy aliens who

After being searched, Banff

were in violation of the law and to detain them as civilian prisoners. It was, however, beyond the scope of the powers of the government to intern as well as treat enemy aliens as de facto *military* prisoners of war. Indeed, in international law a clear distinction was made between combatants and non-combatants – article 3 of the Hague Convention was most explicit in this matter. The distinction was important, because it identified the basic criterion determining who could be considered and treated as conventional prisoners of war.[15] For the vast majority who were interned as prisoners of war, this distinction was rarely made, and, more fundamentally,

"I Do Not Think That Canada Would Take Their Own People"

I, Jacob Kondro, received a letter from my wife that my son, John Kondro, is in prison at the internment camp, Banff. I do not think that Canada would take their own people and put them in prison in an internment camp. I am naturalized as a citizen of the Dominion of Canada. Please let him go.

Jacob Kondro, Dalmuir, Alberta
to Gen. E.A. Cruikshank, Commanding Officer, Military District No. 13
8 February 1916

I have the honour, in reply to your letter regarding the marginally noted Prisoner of War [John Kondro, No. 224], to state that this man claims he is only seventeen years of age and has been in this country for eight years living with his father. He also states that his father is a naturalized British subject, but he himself has not taken out naturalization papers.

I have looked up his record and find that, while Major Stuart was here, he received several days detention for refusing to go to work. However, I think he was led into this by one of our old troublemakers. Except for the above mentioned instance, his conduct has been good and he has been a good worker. It would appear to me that if his father was naturalized before the boy became of age, he himself becomes a British subject, therefore it would not be necessary for him to take out naturalization papers.

Capt. P. M. Spence, Commanding Officer, Banff Internment Camp
to Gen. E.A. Cruikshank, Commanding Officer, Military District No. 13
17 February 1916

National Archives of Canada
Record Group 24, vol. 4721, file 2

it was never used to determine who should be interned and compelled to work as a de facto military prisoner of war.

As for the statement made by E. Newcombe, the deputy minister of justice, declaring that neither international practice nor international law prevented Canada from employing the labour of foreigners who had been cast upon the charity of the state, it was misleading in that it failed to acknowledge the nature of the problem. The issue was not whether Canadian officials had the legal authority to intern enemy aliens who were in violation of the law or to have prisoners of war work for their own keep or otherwise. The point was that the government of Canada was outside the law when, by means of a legislative decree, it targeted enemy aliens as one class among the thousands of indigents and when it did so in order to arrest and imprison them and force them to work as military prisoners. Despite the enemy appellation, resident aliens had rights and could neither be interned as military prisoners (unless they were reservists) nor forced to work as such. This is to say nothing, of course, of the government's responsibility to meet the provisions in international law entitling enemy aliens to notice and to sufficient time to quit the country if they so desired. Canadian authorities were not prepared to live up to this responsibility, and, in fact, they took steps to prevent aliens who wished to leave from doing so.

In the end, a few elected officials did sense that the measures used against enemy aliens – registration, internment, or, as in 1917, disfranchisement of tens of thousands – violated the essential spirit of the implicit relationship between Canada and the newcomer.[16] When the challenge of war dominated much of the public debate, it was a rare moment

Compound gates, Jasper

"Do Not Forget of Me"

General! Writing one more petition to you Sir! I am begging you to inform me if I can be released from detention camp. From several weeks my naturalization papers are laying in the office of Major Commander of your camp and now he says that I have to have someone to guarantee for me. If I have my Canadian paper, if I am loyal to this country in which I am from twelve years, I don't think guarantee is necessary.

So, I beg you, Sir General, do not forget of me.

Prisoner Joseph Leskiw, Castle Internment Camp
to Gen. E.A. Cruikshank, Commanding Officer, Military District No. 13

n.d.

National Archives of Canada
Record Group 24, vol. 4721, file 1

indeed when the former prime minister, Sir Wilfrid Laurier, stood in the House of Commons and addressed the assembled members to speak not of honour and glory but of responsibility and promises. Objecting to the government's decision to disfranchise resident aliens just before the wartime election of 1917, he called on his colleagues to reflect upon the deeper meaning and implications of their actions:

> We want more people, we want more help and more assistance, and where do we get it? Do you believe, Sir, that when peace is restored … when we shall send our immigration agents to Europe again, as we did before, to bring immigrants to this country, that we will send them to the Balkans … Do you believe that when our Canadian immigration agents will go to the

> Balkan States [*sic*], among the Galicians, Bukowinians, and Rumanians that these different races will be disposed to come to this country, when they know that Canada has not kept its pledges and promises to the people from foreign countries who have settled in our midst … I am sorry that I have again to dissent with the Government on this measure. But I believe – and we shall be judged some day by our actions here – that the Government is taking a step which will cause serious injury to the country.[17]

Sir Wilfrid Laurier had made a quiet gesture prompted by an intuitive and clear sense of justice and moral right. His admonition, however, had no effect, his words dissipating into thin air in the cavernous hall of the Canadian parliament.

4

The Policy and Practice of Canadian Internment: A Comparative Perspective

BETWEEN SOVEREIGNTY AND POLITICS there is a necessary, if not tense and paradoxical, relationship. Sovereignty grants states authority in domestic matters and preserves their independence in international relations. Indeed, as one of the few political values around which there is a wide, informal consensus, sovereignty has helped to create a system of balance of power between states and has shaped such political ideas as compliance and noninterference in internal affairs, which have enabled the international legal order to evolve. Sovereignty, in effect, serves as the foundation for the modern political order, and it has become, as a result, an important source of political legitimacy. But sovereignty also acts as a catalyst for conflict. To support claims of sovereignty, sovereign rule must also rely on the prospect of using force when necessary, threatening in the process the very order it has helped to create. Hence the paradox and tension that sovereignty represents for politics. The tension is real and palpable and can be resolved only when restrictions are placed on the use of

Rough trail, Banff

force, when power and authority are obliged to operate under certain conditions. Domestically, this means working within a constitution while internationally abiding by treaties and norms. In short, it is insufficient for a state to declare its actions legitimate by simply invoking the principle of sovereign right – the legitimate use of force implies certain obligations that cannot be ignored.

The War Measures Act was predicated on the sovereign right of the state to self-defence. It enabled Canadian authorities to exercise broad executive powers to meet a perceived security threat. Yet the traditions of common law, international public law, and recognized international standards suggest that there were limitations and norms that still had to be met. In the end, then, what would define the legitimate use of the War Measures Act was not the principle of sovereign right but the way in which policy was implemented: whether the state would take into account principles of natural justice and conform to international norms and regulations or whether it would act without regard to convention, custom, or right. During the Great War the issue of how to exercise power

New bridge crossing, Jasper

justly was a major test facing Canada's political leadership. It created an enormous challenge and provided a unique opportunity for a nation in the making.

It could be said, of course, that extraordinary circumstances warrant extraordinary measures, and from this perspective arguments based on international law and international standards may be regarded as academic, having little relevance in the real world. Yet the historical evidence would suggest otherwise. During World War I, internment as a political measure was used by all the front-line European states – a not entirely unexpected development in an age of conscription and mass mobilization when allowing able-bodied enemy aliens to return to their country of origin had serious military implications. Nevertheless, a close examination of the history of internment during the war reveals several curious twists.

For one, the warring European states attempted to work out among themselves reciprocal agreements for the release of their internees, including males under and over the ages of sixteen and fifty-five respectively, while in several cases successful bilateral arrangements were negotiated, leading to an altogether lenient policy toward civilian prisoners. This was true, for example, of the settlements concluded between France and Germany, Germany and Italy, Britain and Austria, and Austria and France.[1] More telling were the actions of Japan and the United States, where, removed as they were from the European theatre of war, the policy of interning civilian enemy aliens had no role to play in the former and only a nominal and highly restricted role in the latter.[2] These examples illustrate that internment policy was "context sensitive" and that the belligerents did try to distinguish, at least in their treatment of prisoners, between genuine combatants and civilian enemy aliens.

There was, in effect, a certain sensibility concerning the rights and responsibilities owed to the enemy alien that the parties fundamentally acknowledged and attempted to accommodate. Although the extenuating circumstances of war dictated caution, the belligerent front-line states implicitly understood that interned enemy aliens were not prisoners of war in the conventional sense of the term, and they sought to mitigate the needless suffering of the innocent. Circumstances had made them pawns of war and, consequently, the various parties attempted to negotiate, even under the most trying of circumstances, settlements that would see their release or lessen the burden of their incarceration.

Canada, however, did not conform entirely to this perspective or practice. Although the war was just as distant and the security ramifications just as remote, Canada, unlike the United States, pursued a policy that interned resident enemy aliens and, more importantly, treated them as captive military prisoners. At the time, of course, much was said about international law and the need to subscribe to convention and practice. Several perfunctory statements were even made in Parliament about not stooping to the level of "the barbarism of the Hun." But the evidence would also show that there was little consideration given to the consequences for civilian enemy aliens who were interned and treated as military prisoners of war and who would languish behind Canadian barbed wire.

A few officials did give some thought to the matter, if only because they were responsible for the operations side of internment, and there were some misgivings over the policy as it was shaping up. General Otter, for example, voiced concern over the reasons behind the arrest of civilian enemy aliens. But he was a professional soldier, and the reasons were a political matter. It was not for him to question the policy but to carry it

"I Have a Wife and Family to Look After"

I beg to ask you for my release again. Quite a number of prisoners of war are being released at present and I would ask you to liberate me together with them. I have a wife and family to look after and they are in very bad need of help now.

Hoping that you will grant this favour for me.

Prisoner Philip Marchuk
to Major P. Spence, Commanding Officer, Banff Internment Camp
22 November 1916

National Archives of Canada
Record Group 6 HI, vol. 759, file 3565

Prisoners and guards, Monashee

"Anxious to Get Some Position"

I received a letter this morning from my brother, Barnard Stewart, 1441 22nd Avenue, Calgary, Alberta, asking me to use my influence in securing for him a commission in a camp that they are starting in Banff. He does not make it very clear to me what he wants in this regard, but I have no doubt your Department will know, since you know what the intention is of starting a camp. All I need to say is that he spent, perhaps, 25 or 30 years on the Toronto Police Force, and, I think, will be a good man for work of this kind. I am writing you with a view to seeing whether you can do anything for him, because, like all others, he is anxious to get some position.

T.J. Stewart, MP
to Gen. Sir Sam Hughes, Minister of Militia
18 December 1915

With reference to your private letter dated 20th ultimo addressed to Colonel A.T. Ogilvie, District Officer Commanding, Military District No. 11, which has been passed on to me along with its enclosure – a copy of a letter received from T.J. Stewart, MP, regarding his brother. I have the honour to state that I have seen Mr Barnard Stewart. He is fifty-four years of age and afflicted with hernia which compels him to wear a truss. He is, therefore, not medically fit for service. At present, I am not aware of any vacancy at the Banff Internment Camp, but if there should be in the future any opportunity to employ him, I shall be pleased to arrange to do so.

Gen. E.A. Cruikshank, Commanding Officer, Military District No. 13,
to Sir Sam Hughes, Minister of Militia
5 January 1916

National Archives of Canada
Record Group 24, vol. 4721, file 2

out in accordance with the military provisions that guided his professional conduct and duty. He therefore attempted to put in place a system of detention that met the standards set out in the conventions governing modern land war, including the treatment of military prisoners. Equally, he sought to maintain discipline among the troops and order among the prisoners, issuing copies of excerpts from the Royal Warrants to each camp commandant and outlining precisely the routine and duties in the administration of the camp and the expected responsibilities of soldier and prisoner alike.[3]

Otter aspired to make internment operations credible. But despite his good intentions, several factors conspired against this end. At the most basic level, qualified personnel were not available for the job. The more able officers were sent overseas, while those who remained or had returned from Flanders at times proved less than satisfactory. Some had enlisted simply to acquire the prestige and status that would accompany an officer's rank. Others had used political connections to gain appointments and easy promotions, either for themselves or for their family members. Unable entirely to prevent this practice, Otter would be saddled with staff of questionable character and ability.[4] Neither was he able to entirely control the quality of the guards assigned from the Militia Department, the agency responsible for providing personnel. Because of the Directorate of Internment Operation's low priority in the war effort, its ranks were soon filled with individuals who, for one reason or another, were not suited to the work. The overall result was uneven administration of the camps. Abuse of power became commonplace, and many liberties were taken with the prisoners, leading to a less than satisfactory record of performance.[5]

The shortcomings of Canadian internment operations, however, were not strictly speaking entirely associated with the inadequacies of personnel. Rather, the ambiguous status of the enemy alien as prisoner of war helped to create a climate that enabled violations to occur. The transgressions were, to be sure, in an immediate sense a function of neglect, frustration, or zealousness on the part of several officers and members of the rank and file. But because the status of the enemy alien was uncertain, the norms defining duty and responsibility were neither altogether respected nor necessarily observed.

The basic problem was that interned civilian enemy aliens were not military prisoners of war, something well understood by several officials charged with the welfare of the internees. Yet the work expected of them depended on their being designated prisoners of war as conventionally understood. This anomalous situation led to two tendencies that, when combined, effectively made the position of the interned civilian enemy alien untenable. Since interned enemy aliens were in a real sense noncombatants and generally recognized as such, the rights guaranteed military prisoners of war under the Hague Regulations would not apply to them. On the other hand, having been designated prisoners of war, they would still be subject to a work regime that went well beyond accepted and normal practice in the treatment of interned civilian enemy aliens elsewhere.[6] The legally ambiguous position of the interned enemy alien – neither civilian nor military prisoner – allowed a policy of systematic exploitation to take hold. This feature of internment policy would distinguish Canada from the other warring parties, a conclusion that is best understood in a comparative context.

Throughout much of the history of international law, from the period when the first principles of "civilized warfare" were being laid out by the great seventeenth-century Dutch jurist Hugo Grotius, there has been a debate about who should be considered a prisoner of war. At a most basic level and in the context of organized conflict among states, it was the enemy lawfully bearing arms and captured during the course of war who was liable to be interned and treated as such. The conventions of war would hold that, at a minimum, the principal criterion determining prisoner of war status was direct and formal involvement in hostilities. Civilians who were not directly party to the conflict and who were regarded as noncombatants were considered to be in a position of neutrality and, consequently, were not to be interfered with.[7] The nature of twentieth-century war would change all that, shaping the issue in such a way that civilians – and therefore civilian enemy aliens – could be viewed under certain conditions as potential combatants. This inevitable and logical consequence of modern warfare meant that civilian enemy aliens might possibly be interned.

But could an interned civilian enemy alien be considered a prisoner of war, legally speaking? In the context of the early twentieth century, international law lagged behind the needs of the moment. Wrestling at the time with a problem made real by belligerent states pursuing a policy of systematic internment on the basis of perceived military necessity, international jurists were uncertain in this matter. Even after much speculation and debate the juridical problem would remain unresolved, not to be seriously taken up again until the more egregious aspects of a still later twentieth-century global war made it necessary to adopt international legal norms with respect to this question.

"There are Papers Here"

I beg to acknowledge receipt of yours of the 18th instant and contents noted. In reply, I beg to advise you that your naturalization certificate was granted by the Judge in May, 1913 and on receipt of the sum of 50¢ we will forward you a copy of the same.

Office of the Clerk of the Court, Medicine Hat
to Prisoner Nick Goriuk, Castle Internment Camp
20 July 1915

I have the honour to enclose herewith a letter received by Prisoner of War No. 10, Nick Goriuk, from the Clerk of the Court at Medicine Hat which shows he is a British subject. He is writing for his naturalization certificate and when this arrives I suppose he will be entitled to his release. Will you please instruct me. There are papers here showing he was interned for disorderly conduct, but a British subject cannot be permanently interned for that of course.

Major Duncan Stuart, Commanding Officer, Castle Internment Camp
to Col E.A. Cruikshank, Commanding Officer, Military District No. 13
25 July 1915

I have the honour to forward herewith the naturalization certificate of the prisoner marginally noted who had applied for release. In this connection, I beg to state that this man was interned for "Disorderly Conduct" and request that I be advised of your decision in this matter and that the naturalization certificate be returned.

Col E.A. Cruikshank, Commanding Officer, Military District No. 13
to Major General Wm Otter, Director, Internment Operations
5 August 1915

National Archives of Canada
Record Group 24, vol. 4729, file 3

On the other hand, for many of the states participating in the Great War, the demands of the conflict as it was unfolding were real and pressing, and the status of the enemy alien in international law was of less concern than the sovereign right to intern combatants, perceived or real. To be sure, exercising that right was problematic insofar as the juridical status of the enemy alien remained ambiguous, but the problem could be resolved as long as the enemy alien was thought to be a potential combatant. In large measure then, the detention of enemy aliens was predicated on the need to view them as bona fide combatants. Whether they were combatants in the real sense of the term was largely irrelevant. What was important was the perception or conviction that they could be.

Tolerating the idea that in modern warfare there were *potential* combatants, at least opened the way, even if it did not make it clear, to the possibility of interning civilian enemy aliens as prisoners of war. The idea was crucial because the issue of the status of civilian enemy aliens as prisoners of war now no longer mattered at a certain level. They were simply prisoners of war.[8] The designation was useful because it enabled national authorities of belligerent states to deal cleanly with the legal issues that arose from the peculiar predicament of interning people who were in effect civilians. And yet the prisoner of war classification did not release these authorities from other legal obligations. On the contrary, interned civilian enemy aliens designated as prisoners of war were entitled to certain undisputed prisoners' rights that were widely recognized under the international rules of engagement. Therefore, the Canadian experience must be assessed in the light of several pertinent questions: were the prisoners' rights as prisoners of war being observed and were the necessary

Wounded prisoner, Castle Mountain

steps being taken to ensure that the international standards to which Canada was both legally and morally bound were not being violated?

The Hague convention of 1899 and the amendments to the convention of 1907 governing the treatment of prisoners of war were explicit about their rights.[9] The key articles of the agreements proscribed the use of prisoners in excessive, dangerous, or humiliating work. Furthermore, although the state could authorize and expect to use prisoner labour for public service as long as it had no connection

"What Did You Think the Soldiers Had Rifles For?"

I have been a prisoner of war at the Castle Camp since the 1st of August. I tried to escape on Saturday 12th August with two other prisoners of war. We arranged to escape on Friday. Prisoner of War No. 290 proposed it to me and said that the fellows who escaped before had a chance and we might try. I took some bread in my pockets. We were sitting – resting – and when we were told to start work again No. 290 said in our own language [Ukrainian], "Now we're near the bush, let us try here." We made a dash for the bush and we were quite a way in before we heard "Halt" and about twenty-four shots went off. I was lying in the bush at the time the shots went off, taking cover. I was a soldier in the 90th Battalion CEF [Canadian Expeditionary Force] and was transferred to the 190th when the 90th went overseas. I was a soldier for eight months. After the soldiers stopped shooting, I got up again and went west. As one of us was shot and the other fellow did not want to go on, I thought I would go along to the steam shovel gang and give myself up. As I was passing the engineer's camp, a teamster saw me and asked me where I was going.

I told him "we were beating it" and that one of us was shot and the other did not want to go further while I was going to give myself up to the steam shovel party. He said, "You can't go further, wait here until the soldiers come." I remained there until the soldiers took me in charge.

EXAMINED BY THE COURT

Q: When you were arranging this escape did you not think there was a danger of your being shot?

A: I did not think about that.

Q: What did you think the soldiers had rifles for?

A: No. 290 told me they don't shoot at the men. They just shoot in the air, or in the ground, and then run after you.

Q: Did any shots come near you?

A: No sir. I was laying on the ground and the shots went far away over me.

Q: Do you think that you'll risk it again?

A: No sir, I will not.

Testimony of Prisoner of War No. 598 John Dykun
Board of Inquiry, Konowalczuk Shooting
21 August 1916

National Archives of Canada
Record Group 24, vol. 4728, file 2

with the operations of the war, it was to be compensated for with pay commensurate with "rates in force for work of a similar kind done by soldiers of the national army." Underpinning much of this was the working principle that through the act of acquiescence, by "laying down their arms," prisoners of war recognized the authority of their captors. But in the process, responsibility for their security and well-being was also conferred on the state to which they submitted.

The relationship between captor and prisoner was in this sense infused with the spirit of civilized behaviour, while the justice in the relationship was defined by a mutual acceptance of the roles and responsibilities that lay with each. Prisoners of war were expected to comply with the rules of conduct; if they did not they could be disciplined. The captor, for his part, was to treat prisoners humanely, providing them with the same standard of welfare – food, shelter, and clothing – as was enjoyed by his own men. Prisoners would be allowed to keep their personal belongings, and relief societies were to be assisted in performing humanitarian work. As for disciplinary punishment, when applied, it was to conform to procedures outlined in universally recognized codes of military conduct; physical violence to enforce obedience was contrary to the meaning of the term "humane" and expressly forbidden. Finally, at the conclusion of hostilities all prisoners of war were to be repatriated as quickly as possible.

The Hague agreements rested on the direction provided by a growing body of prior agreements, primarily bilateral arrangements, that sought to affix the proper conduct of warring states with respect to captured men-at-arms, guiding, for example, Germany and Britain in their negotiations on the treatment of POWs during the most difficult years of the Great

War.[10] In keeping with the Hague conventions the Anglo-German agreement on the treatment of prisoners that was finally concluded in 1917 not only reaffirmed the rights identified in the earlier conventions but also made provisions that protected prisoners from violence, personal insults, and public curiosity; established work norms that restricted the work day to ten hours (including travel to and from the work site); and prohibited the use of prisoners in certain types of arduous labour if physically unfit or inexperienced by virtue of their previous occupations.[11] In the final analysis, however, these gestures were motivated less by a sense of altruism than by the practical consideration of mutual interest. Contravention or abridgement of the guidelines could lead to retribution and increase unnecessary suffering. This was to be avoided if at all possible.

Modern warfare, of course, made compliance difficult to observe. The employment of prisoners of war in any form of work directly augmented the fighting capacity of the captor by releasing home labourers. Consequently, strict observance of the law proscribing the use of prisoners of war on projects that had some bearing on military operations was moot in the context of nations at war. Yet, curiously, the belligerents were careful in interpreting this and other clauses outlining POW treatment.

Prisoners, for instance, were required only to take care of themselves and their premises, and although they could be authorized to work and compensated for it, they were theoretically not compelled to do so. From 1914 to 1916, only a small percentage of the prisoners of the warring states were in fact working. As of January 1915, in France a scarce 10 percent of prisoners were employed, while in Britain, only 7,000 of an estimated 150,000 POWs were working as of March 1917.[12] This would change in the

Road work, Mara Lake

final years of the conflict, when all the European belligerents resorted to the desperate strategy of imposing economic blockades on their enemies, leading to severe food shortages. Prisoner labour was consequently conscripted and widely used in agriculture. In 1917 and 1918, for example, 900,000 Imperial Russian prisoners of war were employed sowing and harvesting crops in Germany and another million in Austria-Hungary. In Britain during the submarine blockade of late 1917, German prisoners were put to work on marginal lands, resulting in a marked increase in prisoner labour at this time.[13] Importantly, however, for most of the belligerents, the use of POW labour was a response to the economic blockades that had been imposed; it was not conceived of as a policy measure in its own right.[14]

Where POWs were being used, the belligerents were careful to avoid the impression that their labour was being exploited. Although international law provided for the authorized use of POW labour, it was to be unrelated to the war effort and compensated for, and it was not to be either excessive or prejudicial to the health of the prisoners. The use of force as a means of extracting work from prisoners was condemned and forbidden. Conditions, of course, varied from country to country and situation to situation. That abuse was prevalent at the local level, there is no question, and indeed there is considerable evidence of it. But as *principles*, these elements of international law were generally recognized, and an attempt was made to follow them.[15] Prisoners of war were to be considered not as slaves but as vanquished soldiers. Indeed, despite the carnage of the war, the warrior code still applied, and the ethos of clemency and mercy that was very much a part of the European tradition of military chivalry meant that the liberal notions of captivity would be adhered to.[16]

For Canada, on the other hand, the experience of prisoners of war and their treatment was singularly different, in that use of prisoners as a source of labour was a routine practice, not a response to the adverse conditions brought on by war but a crude utilitarian measure. The principle was simple. There was work that needed doing and prisoner labour, which was suddenly available and seen as an opportunity, would be utilized. Indeed, it was felt that internment labour should be put to good effect. And so it was.

As prisoners of war the internees on the Canadian frontier were pressed to work hard and for long hours under primitive conditions.[17] Many were unsuited to the work, having had no previous experience, while several who were suffering from serious ailments were clearly in no position to

"We Cannot Go On Much Longer"

I am glad to have received your welcome letter. I am very glad to hear from you that you are back from hospital and that you are in better health though you say you are very weak. I believe you but I cannot help you. As you know yourself, there are men running away from here every day. The conditions here are very poor, so that we cannot go on much longer. We are not getting enough to eat – we are hungry as dogs. They are sending us to work, as they don't believe us, and we are very weak. Things are not good. The weather has changed for some time past and it is wet and muddy. Also in the tents in which we sleep, everything is wet. We get up at 5 o'clock in the morning and work till 10 o'clock at night. Such conditions we have here in Canada, I will never forget. Men have escaped from here – 28 now.

Nick Mudry ran away yesterday. You might tell his wife. But I must wait till the end because I have been here 10½ months already. I don't wish to lose the money I have earned here. My dear wife, please try to find somebody to help you because you are not able to go to work. I am sure you are very weak, and I would advise you to write a letter to the Camp Commandant asking for support. If they refuse to give it to you, ask them to release me so I could support you as you need. I have nothing else to write you, only to wish you better health.

Censored letter, Prisoner N. Olynik, Castle Internment Camp
[28 October 1915]

National Archives of Canada
Record Group 24, vol. 4729, file 3

undertake manual labour.[18] Food was scarce and wanting and the work difficult, exhausting, and dangerous.[19] Those who refused to work were deprived of food, a coercive measure that was widely used both in extracting work from the internees and in managing troublesome characters and camps. The conditions were variously protested, but with mixed results. Invariably the complaints were trivialized or dismissed. Humiliated and taunted, many prisoners simply resigned themselves to their miserable existence. For those who resisted, corporal punishment and other disciplinary methods were used to achieve compliance. Unprovoked physical violence and personal abuse from undisciplined guards who resented both their work and the prisoners they felt were the reason for their lot were more common.[20] The frustration was so pervasive that large numbers of guards, still motivated by the beliefs that had initially inspired them to volunteer, asked to be transferred overseas, fearing that they would be saddled with the ignominious reputation of a prison warder whose only contribution to the Great War had been overseeing numerous unfortunates who had the bad luck of being enemy aliens and unemployed.[21] The experience would prove troubling for some guards and add to the melancholia of a few lost souls who were already depressed and who would in desperation take their own lives.

It was a distressing situation all around. Much, of course, was accomplished, but more often than not as the result of a callous, if not cruel, disregard for the rules of conduct, something that became manifest even in the most mundane ways. It was not unusual, for example, for internees to be forced to perform menial tasks not only for local commercial interests but also for the personal benefit of officers, guards, and officials alike. In

Enemy aliens in prison compound, Monashee

addition, the cost of provisioning was closely monitored, but the exercise soon went beyond austerity. Whenever possible, food substitutes were introduced to reduce expenses, resulting in both a decline in caloric intake and frequent and widespread complaints of hunger.[22]

Meanwhile, as a result of an effort to bring costs down further, the civilian internees were allotted only twenty five-cents a day – the supplement paid to soldiers for nonmilitary work – rather than the standard soldier's

A Letter

To be known to you that we, prisoners of war, are treated here very bad, most inhumanely, what can be proved facts. When you was here to see us they sent me out of the camp on Sunday so that I could not make any complaints. I, the undersigned prisoner of war, was struck on the face by Major's stenographer for only asking him why I had to sign my name twice and was hit with a bayonet. So, I beg you, please say your word about this.

Intercepted letter, Mike lwasieczko, Banff Internment Camp
to the American Consul, Calgary
n.d.

National Archives of Canada
Record Group 24, vol. 4721, file 1

rate guaranteed under international law. Since the supplemental pay was the only income source by which the simple pleasures of life – tobacco and sugar – could be purchased in the camp military canteens, withholding pay was used as a form of punishment. Significantly, doing so had the added benefit of creating the impression of a healthy financial balance, which flowed naturally from the official preoccupation with keeping operational expenditures to a minimum.[23]

These commonplace irregularities pointed once again to the fundamental ambiguity in the status of the enemy alien as a prisoner of war. For camp administrators it was evident that the internees were not prisoners of war, certainly not in the conventional sense of the term. This was critical because it meant, therefore, that they were unentitled to the rights

Inspecting work on the Spray River bridge, Banff

that were identified under the rules of engagement; a perception reinforced by a division of responsibility for the prisoners. Although the Internment Operations Directorate was entrusted with their confinement and maintenance, agencies to which the prisoners were contracted, such as the Parks Branch and the Department of Public Works in British Columbia, set their work schedule. Disputes over authority were consequently frequent.

Since the contracting agencies were responsible for their prison wage, nonmilitary aspects of the operation increasingly took precedence. For the park foremen and government road superintendents who exercised de facto control in the field, the issue was not security but rather that the internees worked and worked hard. As for the regulations governing prisoners' rights and obligations, they simply complicated matters, and there was persistent pressure to dispense with rules governing prisoner treatment that hindered the progress of work. The problem, of course, was that it was not entirely possible to ignore the prisoner of war designation or the need to appear to be following the regulations, because to do so would undermine the government's perceived legal authority to authorize the work.

The harsh regime to which the prisoners were subjected did not go unnoticed in Berlin. In July 1915, for instance, a *note verbale* protesting the alleged inhumane treatment of German prisoners of war was sent to the British government, which in turn notified its Canadian counterpart. Canadian officials responded by noting that only ten German prisoners were working and this in a voluntary capacity as cooks for other first-class prisoners – the distinction made between officers and enlisted personnel in accordance with the Hague regulations.[24] This reply was astonishing, however, in that it failed to reveal the extent and nature of prisoner labour. Yet the omission was not deliberate. German prisoners – many of whom were either quasi-military prisoners who had been captured as seamen or civilians deemed to be of an "officer" class – were in a manner of speaking treated liberally, certainly in comparison to the treatment accorded their working-class Austro-Hungarian counterparts.[25] But more to the point, a distinction was being made between German and designated Austrian prisoners, a matter of importance because it demonstrated in a real sense who constituted a genuine prisoner of war. The comparatively liberal treatment of German prisoners implied that there was a military standard of conduct that was being followed, but that standard did not in fact apply to Austro-Hungarian prisoners of war. It did not apply, simply because Ukrainians and other "Austrians" were civilian indigents and therefore did not enjoy the rights that, for example, merchant marines as foreign nationals or gentlemen of an "officer" class were thought to possess.[26]

Of Stringent Measures and Inducements

Entire camp on strike this morning due to stringent measures instituted by Lieutenant Brooks. Stop. He is handling the situation in proper way and if left alone will get results we want. Stop. Can you support him for Commandant of this camp. Stop. Captain Harvey away. Stop. Brooks took action in his absence.

J. Black, Road Superintendent, Vernon-Edgewood Road
to J.E. Griffith, Deputy Minister, BC Public Works
29 April 1916

For your information, I can state that Captain Harvey is paying little or no attention to getting the prisoners out. Working with 2 of the non-commissioned officers, Lieutenant Brooks has just finished giving the prisoners a session of 5 days on bread and water. After 3 meals today of full rations, he will put them back on 6 days of bread and water. This is without Harvey's knowledge and it should have the desired effect very shortly. The camp Medical Doctor is also supporting Brooks in his effort to bring them to tune. Yesterday, I overheard Brooks ask Harvey to either go away for a few days, so he (Brooks) might have a free hand, or authorize him to take out a couple of the ring leaders in this trouble and administer to them a good thrashing. Harvey refused. Brooks is determined to carry his point with the Austrians, and if let alone will do so. I understand the officer coming to take Harvey's place will be here in a few days.

J. Black, Road Superintendent, Vernon-Edgewood Road
to J.E. Griffith, Deputy Minister, BC Public Works
15 May 1916

(cont.)
Since advising you on the 15th that the internees still refused to work, there has been no change in the situation, except that, on the 18th, the Officer Commanding received the following wire from General Otter in Ottawa:

> Understand work by civilian prisoners of war not compulsory therefore refusal to work entails no punishment. Stop. Encouragement should be given to work and those refusing, who are without funds, shall be given no canteen privileges whatsoever. Stop.

On receipt of this wire, the strikers were immediately released from confinement. The work gang, since, has averaged 9 men. With such a small force, it is going to take some time to complete the canyon road and open it to traffic in a safe and finished condition. I do not think there is any chance, without some inducement, of getting more of them to work. Would it not be feasible to pick out a gang of the best workers, there being fifty or sixty first class labourers among them, and offer them 50¢ per day, with the distinct understanding that slacking would immediately remove them from the road? I am not sure that this scheme would work, but I would like permission to try it out. If the information quoted in the wire becomes generally known to all the internees, it will mean nothing but trouble and wasted money to replace these internees with others.

J. Black, Road Superintendent
to J.E. Griffith, Deputy Minister
23 May 1916

Referring to your communication of the 23rd instant, I beg to advise that we have no intention of paying the aliens 50 cents a day.

J.E. Griffith, deputy minister,
to J. Black, Road Superintendent, Vernon-Edgewood Road
29 May 1916

BC Ministry of Transportation
Okanagan District (1916), file 1752, section 4

The view that the interned Ukrainians and other East Central Europeans were not prisoners of war in the conventional sense was shared by neutral observers entrusted with representing the interests of the internees as prisoners. The Calgary-based American consular official, Harold D. Clum, who was tasked with investigating complaints in the camps located in Alberta's Military District 13, noted that although internment conditions were rigorous, they were not something that the interned, as members of a working underclass, were unaccustomed to. For Clum, the issue was not whether international law was being violated but whether the work schedule was too excessive given their station in life.[27] The ambivalence of the neutral observer echoed in the lack of interest of the Austro-Hungarian government, which made no inquiry, let alone protest, thereby not only highlighting the marginal status of the national minorities in the old empire but also the underlying belief that this was now a domestic and internal Canadian problem. With no mechanism for appeal and little external interest in and support for them in their predicament, and foreseeing no end to the harsh conditions of their imprisonment, it was not surprising that more than a few internees risked being shot escaping to avoid having to face more of the same.[28] The majority, however, fell into a deep despair when they found themselves in a totally incomprehensible situation.

It was incomprehensible to them because they at least understood that they were not prisoners of war. They were not party to the conflict: they were simply civilian aliens who had the misfortune of being interned. There was a qualitative difference between the two categories and, unlike Canada, the European and other warring states had no trouble making

"In All Degrees and Stages of Raggedness"

I have the honour to inform you that I have received a letter from Major Duncan Stuart commanding the internment camp at Castle, Alta., which reads as follows:

> I am sorry that so many prisoners have got away but except in one or two cases I cannot blame myself. They are without boots in numerous cases, that is good boots, and nearly, if not quite half of them are wearing overalls in all degrees and stages of raggedness. Some are going about in under-drawers only and were until yesterday sleeping in tents without fires. I have asked the Office in Ottawa three times by wire to let me buy some boots and overalls locally but they probably think it better to wait until the regular shipments arrive. I am sending forward an official letter today showing what has been done to get the prisoners into heated tents.

I intend to proceed to Castle at an early date and will report more fully on my return to Calgary.

Col E.A. Cruikshank, Commanding Officer, Military District No. 13
to Gen. Wm Otter, Director, Internment Operations
5 November 1915

National Archives of Canada
Record Group 24, vol. 4729, file 3

Road work, Edgewood

that distinction. It was premised on the official attitude among the belligerents that they had no specific quarrel with civilian enemy aliens, a notion that followed from the principles laid down at the Second Hague Peace Conference, which, for example, made it clear that it was "considered inadmissible to imprison subjects of the enemy who at the outbreak of the war are on the territory of belligerents."[29] Enemy aliens, of course, were interned in Europe, but more as a function of events borne of suspicion and rivalry and not as a calculated move. More importantly, their internment was not predicated on a policy of coerced labour, a conclusion that requires closer examination.

"Causes Unknown"

On Sunday afternoon, December 24th, between 3 PM and 4 PM, I was called to the internment camp, Banff, Alberta to see a prisoner of war named George Luka Budak, who, it was alleged, had cut himself with a razor.

The man's abdomen was fearfully gashed and the bowels were hanging out. There was also a deep gash across the throat. Dr Poyntz, who was with me, took charge of the case, but the man died in about an hour. As this occurred on a Sunday, and the following day was Christmas Day, I empanelled a jury to view the remains Sunday evening. I then had the body removed by the undertaker. The jury adjourned until 3 PM, December 26th, when the inquest was proceeded with and the evidence taken which is herewith enclosed. The jury required but a few minutes deliberation to reach the following verdict: *"The deceased came to his death by wounds self-inflicted – causes unknown."*

Verdict of R.H. Brett, Coroner, George Luka Budak Inquest
27 December 1916

Public Archives of Alberta
Acc. No. 67.172/947

With war imminent, it was unclear what policy the belligerent states would follow towards the mass of foreigners located on their respective territories. Of all the European states, France was in the most difficult position, playing host to some one and a half million immigrants, the majority being enemy subjects. However, in a spirit of goodwill, a short grace period enabling foreign nationals to leave the country was provided. In Britain and Germany, where the numbers of enemy aliens were significantly lower and the problem of their disposition less pressing, a longer period was granted. As for Austria-Hungary, there were frequent reports of enemy aliens leaving at their leisure well after hostilities commenced. It was not until the scope of the conflict and the potential for national calamity was fully understood by the belligerents that the drastic measure of internment was implemented.

In France, the initial catalyst was the early success of the German army. Paris, with its 400,000 enemy alien inhabitants, was on the verge of being seized, and in the hysteria of the moment enemy spies were thought to be everywhere. Meanwhile, the sinking of the *Lusitania* in 1915, with its attendant loss of civilian life, proved to be a turning point in British policy. Until then, Britain had relied primarily on emergency legislation to restrict the movement of aliens and prevent acts of sabotage. Public outcry, however, over the loss of innocents on the torpedoed passenger ship demanded nothing less than retribution. The wholesale internment of enemy aliens in Britain was consequently seen as practical and logical.[30] The British response, not surprisingly, resulted in reciprocal action on the part of German authorities, who felt compelled to reply in kind.

Prisoners at Squall Mountain, Banff

Enemy aliens had become a complicating factor in the conduct of war. Although it was conceded, over time and generally, that they did not constitute an actual threat, the ever-expanding repertoire of twentieth-century warfare meant that internment would be resorted to as a legitimate instrument of war. But the imprisonment of enemy aliens masked an element of uncertainty in the use of the measure. Germany, for instance, initially proposed a universal exchange of enemy aliens, without exception for age or category. Britain, however, rejected the idea, since the disproportionate numbers held in the United Kingdom would have given Germany an unfair advantage. The problem of enemy aliens would not be easily resolved. But if they did not represent a legitimate threat, then how were interned enemy aliens to be treated? This awkward situation required a sympathetic understanding of the circumstances and, correspondingly, some accommodation.

The treatment of interned civilian enemy aliens among the European belligerents, as well as in the United States, which had entered the war in 1917, was nothing short of remarkable, given the scale and intensity of the conflict. Because whole societies and economies were geared to achieving final victory, the welfare of enemy aliens should have been of little concern to political and military leaders interested principally in the successful prosecution of the war. Yet every effort was made to ensure that their internment would be neither harsh nor penalizing.[31] Indeed, although detention was an unwelcome and trying experience, the authorities were careful to ensure that there was as little discomfort as possible.

Camps, accordingly, whether in Germany, France, Britain, or the United States, were organized and administered to provide civilian inmates with considerable autonomy. Moreover, since enemy aliens were not military prisoners, there was no obligation for them to work beyond what was necessary to keep the camps clean and orderly. Consequently, much time was spent by the internees on creative endeavours and intellectual pursuits. Dramatic, musical, and horticultural societies flourished, as did reading and learning circles. Recreational sport was also encouraged to ensure fitness among inmates. Meanwhile, hygiene, as well as physical and mental health, were so important that in an astonishing gesture of compassion and goodwill, detainees in France were allowed to leave the camps twice daily for walks upwards of one hour and at a distance of one kilometre, in order to alleviate the pressures of captivity. There was overcrowding, to be sure, and there were problems associated with sustenance, especially after the imposition of economic blockades at the conclusion of the war. Trying psychological and practical problems were also associated with this senseless imprisonment. But if there was a chief complaint among many of the civilian internees, it was about boredom.[32]

By all accounts, in comparison to the conditions faced by military prisoners, the treatment of civilian internees in Europe and the United States was liberal, in part because of a general empathy for the plight of these unfortunates, who had the bad luck to be victims of an unpredictable diplomacy. But more to the point, it was acknowledged that the individuals in question were not without rights and that the obligations and responsibilities that followed had to be met. Indeed, a significant number of the incarcerated were domiciled aliens, resident immigrants who had settled permanently in the host country. Residency, although not a guarantee, did entitle immigrants of enemy nationality, despite the extenuating circumstances of war, to some degree of protection under the law, as well as to recourse before the courts.[33] It was recognized that they happened to be on the wrong side of the barbed wire fence only because they lacked citizenship in the unusual circumstances of total war. Natural justice, therefore, warranted that consideration be given to their predicament and that their legal standing as quasi-members of a community in possession of certain rights be recognized.

The treatment of interned civilian enemy aliens in Canada differed markedly from that of their counterparts in other warring countries, the most distinct feature being the systematic and compulsory use of civilian internment labour. There were, of course, some parallels. Conditions at the camps and receiving stations in urban settings, at Vernon, Fort Henry, and Halifax, for example, which housed enemy aliens of German nationality, or so-called "first class prisoners of war," approximated conditions in

Europe. Moreover, unlike the regime in frontier camps, work was not required in these camps. Consequently, as in Europe, beyond keeping the camp orderly and clean, internees devoted themselves to recreational and educational activities.

Significantly, the two situations were similar because, first of all, the authorities were to follow the distinction in the Hague convention between first- and second-class prisoners of war. But critically, the politicized nature of the more worldly German inmates weighed on Canadian authorities and led them to respect and follow international practice in these matters. Indeed, because German internees maintained close contacts with the various consular officials representing German interests in Canada, Canadian officials were careful to oblige. After 1917, when German prisoners were integrated with other detainees of enemy nationality as a result of the need to consolidate camps, their influence became more pronounced as "Austrian" internees, taking their cue from their German counterparts, became increasingly militant. By agitating for and demand-

Twenty-five degrees below, Banff

Suffering Morally

We received your different letters at the camp, and it was most surprising to us, that you, Honourable Sir, in nearly every letter, urged us to strictly obey all the regulations of the camp. As you have pointed this out in nearly every letter, we have come to the conclusion that somebody must have given you a wrong picture of us. You, Honourable Sir, must have the impression that we are outlaws, men who have never obeyed orders, or men who have never thought, even before the war, to live as law-abiding people. But, thanks to the education which the Imperial Government gives to her subjects, your impression is not the right one. The biggest majority of the prisoners here never had anything to do with the police before the war started and never had been inside a prison before. We can earnestly say that we, as German citizens, are well accustomed to obeying the law as far as the law does not make us slaves. But a law which is based upon slavery, we, of course, cannot consent to …

We are not troublemakers and we would welcome a way which would be conducive to a better understanding between officers and prisoners. But to force us to do what we call dishonourable work would not be the right way to a better understanding. Quite a few of us suffer morally by being forced to do such work, and quite a few have given proof that they would sooner take any punishment than to be treated like slaves. Instead of being forced to do disgraceful work, ours is a fate to be pitied. Of course, there are some prisoners who don't feel that way, and we will not bother them or keep them away from work. But prisoners who suffer morally by doing such work should be left alone. In fact, since you were up here, Captain Mitchell has not forced anybody to work, but he has told one of us that we can be forced to work. It cannot be the intention of the Imperial German Government that we civilian prisoners be forced to do disgraceful work. There must be an agreement between the nations at war concerning the treatment of civilian prisoners. According to such an agreement, we want to be treated under the laws which the Imperial German Government has given her consent. This is the only protection we have, as we know, if we are not treated according to such law, we can report it to you, Honourable Sir, as our protector, for transmission to our government and our government has the will and means to protect us …

(cont.)

Of course, we do not want that British civilian prisoners in Germany should be treated badly, as we are all human beings and these poor men feel it just as bad to be in prison as we do. But this is no excuse why we Germans here should be treated like slaves. If you, Honourable Sir, would have been imprisoned for nearly three years, like some of us, you would understand this much better. You would have seen things which you think are impossible or incredible. We have enough proof of it here …

The first thing we need is protection and the only protection we can get is from our government. We are human beings. For the sake of humanity, we kindly ask you, Honourable Sir, as our protector, to send us a written statement, how the Imperial German Government expects that we should be treated. This is the only law which can give us protection.

Secret Letter from Prisoners, Morrissey Internment Camp
to S. Gintzburger, Swiss Consul
5 October 1917

National Archives of Canada
Record Group 6 HI, vol. 765, file 5294

ing Canadian compliance with international norms, they would change the character of the remaining frontier camps, including Morrissey in British Columbia.

By early 1918, with authorities mindful of the wider political repercussions of noncompliance, work at the remaining camps in Canada assumed a more voluntary character. But 1918 was not 1914. And before 1918 the practice in Canada was for civilian enemy aliens to be treated as military prisoners of war. Yet that designation was not without some legal obligations. The laws governing modern land war imposed conditions on

"For the Sake of Humanity"

We beg to inform you that our fellow prisoners in the internment camp at Morrissey are exposed to the very brutal treatment of the Canadian soldiers and non-commissioned officers doing police duty there. As we know that it is impossible for them to write to you stating the facts, and that complaints to the Officer Commanding result only in making things worse for them, we must request you, for the sake of humanity, to inform the Royal Consulate of Sweden, the legal representative of our interests in Canada, of these conditions. We beg him to exercise his influence to have a stop put to the cruel treatment of the prisoners of war at Morrissey.

Prisoners' Complaint, Vernon Prisoner of War Camp Committee
19 April 1918

National Archives of Canada
Record Group 6 HI, vol. 763, file 4738

the behaviour of states at war. But Canadian authorities demonstrated little misgiving in deploying prisoner of war labour because of their unwillingness to extend recognition of POW rights to civilian enemy aliens; they were noncombatants, after all.

But if they were noncombatants, were they not entitled to other rights bearing on their predicament? Sadly, international law was silent on the specific treatment of interned enemy aliens, and within this context very little reservation was shown in using the internees as seen fit. And yet the lacunae in international jurisprudence did not mean an absence of norms. On the contrary, the practice among the warring states was to exercise caution in the treatment of these individuals.

This liberal treatment was in some respects a matter of reciprocity. But this in no way explains the generosity of spirit shown civilian prisoners of war among the majority of the belligerents. Rather, it was generally recognized that precisely because of their noncombatant status, interned enemy aliens were considered more than prisoners of war and as such entitled to greater rights, rather than less. There was also, however, every sense that a certain responsibility, if not obligation, existed with respect to those who chose to reside within the national community, including the alien of enemy nationality. That Canadian authorities did not act in keeping with this understanding, despite the moral and practical imperatives, pointed to the shortcomings of a leadership that failed to meet the deep challenges brought on by war and crisis. For Canada, it was an unfortunate beginning to the twentieth century and an inauspicious introduction to the modern age.

Prisoners and guard, Otter

5

The Prize of War: Internment in the Canadian Rockies

WRITING IN THE INTERIOR DEPARTMENT'S annual report for 1916, Commissioner J.B. Harkin of Dominion Parks equated patriotism with the love of nature: it was nature – "the quality of their country in their own eyes" – that inspired the achievements of Canadians in Flanders. Furthermore, it was this connection with nature, Harkin argued, that would make Canadians "sturdy and rugged citizens."[1] The mountains for Harkin were a tableau and a metaphor for the nation. Pure and simple, steady and solid, they were a measure of Canada and its future.

It was perhaps for this reason that the country's continuing commitment to the development of the national parks in the mountain hinterland was in Harkin's estimation of no less national importance than the battles being fought in distant lands. Harnessing the wilderness so that it might inspire Canadians was foremost in J.B. Harkin's mind. But with the war on and with its competing claims on the public purse, the dilemma for parks officials – and for provincial governments in the Canadian West

Raising the colours, Otter

who were still very much committed to moving forward on public work projects – was how to undertake improvements in a climate of reduced appropriations. The answer would come from an unexpected quarter.

After a registration system for civilian enemy aliens was put in place in Western Canada, the number of detainee-prisoners grew rapidly. To accommodate the numbers, the main holding facilities in the urban centres of Vernon, Brandon, and Lethbridge were operating at capacity, forcing the Internment Directorate to look for additional space. In this challenging situation, General Otter, the director of internment operations, who was obliged to keep costs at a minimum, appeared remarkably well adapted. Recognizing that there was a hidden opportunity in making available prisoner of war labour, he negotiated early on with the Militia Department for permission to use the military installations at Petawawa, Ontario, and Valcartier, Quebec, in return for upgrading the facilities there. Similarly, land grants at Kapuskasing and Spirit Lake that were to be used as sites for prisoner of war camps were acquired in December 1914 after provincial authorities determined that with internment labour these remote areas could be developed for future settlement at little cost.[2] Co-operation was a logical approach to both reducing costs and resolving the predicament of securing more sites. In the early months of 1915 this formula was put to good effect in the West, where the desire to use what amounted to an inexpensive source of labour ensured that there would be no shortage of partners.

Among the interested parties was the government of British Columbia. Widespread anti-immigrant sentiment and an obliging provincial government had resulted in the quick detention of hundreds of enemy aliens.[3] The government buildings at Vernon and at the smaller Nanaimo station were soon overcrowded, leading provincial authorities in the spring of 1915 to consider and accept a proposal to put internees to work on road-building projects, especially on roads through the mountains.[4] The authorities were anxious to improve road access through the Monashee and Columbia Mountain ranges of the Canadian Rockies, and after negotiations had been concluded with the Internment Directorate, a small contingent of interned enemy aliens was transferred from Vernon on 2 June 1915 to a makeshift camp in a remote area of the Monashee range. By the end of the month, 201 inmates had been sent from the Vernon station to the new site.[5] The objective of the Monashee Camp, as it

Grubbing on the Vernon-Edgewood Road, Monashee

was called, was to cut a road simultaneously east and west from the base camp, eventually connecting the town of Vernon with the Arrow Lakes, at which point travellers could continue by boat.

The Vernon Road was an ambitious project spurred on by individual settlers, political riding associations, and local boards of trade, who pressed federal members of parliament and representatives in the provincial legislative assembly to act on behalf of local political and commercial interests. But the project in fact recommended itself and needed no political support, the British Columbia interior holding the promise of future agricultural production. Consequently, despite the logistical difficulties involved in supplying the Monashee camp, every effort was made to ensure that the project was supported and that it continued on schedule. It was only after several months of operation with limited success and unsustainable financial losses, due to both torrential rains and continuing problems of supply, that a decision was made to relocate the camp at Monashee to Mara Lake, British Columbia. News of the closure of the camp triggered a spate of applications from several centres in the belief that they were entitled to their "share of the business resulting from the internment of these men and also any advantage that would come to [them] from public works."[6] All such requests were rejected in favour of a road project at Mara Lake, a difficult and expensive proposition that could, nevertheless, be offset by the abundance of cheap labour.

The closure of Monashee Camp, however, did not mean that the proposed Vernon Road would be abandoned. It was much too valuable. But expenses had to be reduced, and this meant finding a location that could be more easily accessed. With the imminent closure of the Monashee

A Favour

Regarding my interview today on the subject of the employment of alien prisoners at Vernon on the above undertaking [Vernon to Edgewood Road], there now appears to be an opportunity to link up the Okanagan with the Kootenays at little or no expense to the province. Surely this is an opportunity that should be made use of. You were kind enough to promise me earlier your endorsement and active support. I should, therefore, esteem it a personal favour if you would see the Honourable Mr Taylor on the subject and do what you can to help the matter through.

A.D. Ford, Union Club
to Sir Richard McBride, Premier, Government of British Columbia
5 February 1915

I beg to acknowledge receipt of your memo of the 8th instant to which is attached a letter from Mr A.D. Ford, of this city [Victoria], regarding the employment of alien prisoners interned at Vernon on the Edgewood-Vernon Road. May I say that the proposal contained in Mr Ford's letter meets with my strongest approval. I am writing today to Mr R.F. Green, MP, explaining the matter fully to him, and requesting that he see the Minister of Militia to get his views in connection therewith.

Hon. Thomas Taylor, Minister of Public Works
to Sir Richard McBride, BC Premier
10 February 1915

BC Ministry of Transportation
Okanagan District (1915), file: 1752, section 1

"An Excellent Opportunity"

There was considerable agitation last year and considerable pressure brought to bear upon the Department to undertake a road connection running between the Arrow Lakes and Okanagan Lake over Monashee Mountain. On the Vernon end, this road has been completed for a distance of about forty miles passing through an agricultural section, most of which is under development. While on the Arrow Lake side, about seven miles of the road is complete giving access to practically all the developed areas of agricultural land in that section. There remains, however, a distance of about thirty miles to connect the two roads, and the traffic backwards and forwards, at the present time, can only be carried out by horseback or on foot over a very indifferent trail.

I am sure, with your intimate knowledge of this section of the country, you will appreciate that the locality has a decidedly good future from a lake-mining point of view. Notwithstanding the pressure brought to bear last year for an appropriation to connect this portion by wagon road, I found that it was not available. I very much fear that I will not be in a position this year to undertake any portion of this work. It has been suggested, however, by those who are interested in the completion of this road that it would be an excellent opportunity to employ the alien prisoners presently interned in Vernon. May I say that the Premier is very much in favour of the suggestion as are all the local Members [of the Legislature] from Kootenay. No doubt, they will be writing you personally in due course and in connection therewith.

Would you be good enough to take the matter up with the Minister of Militia to find out what his views are with regard to this matter.

Hon. Thomas Taylor, BC Minister of Public Works
to R.F. Green, MP, Ottawa
10 February 1915

BC Ministry of Transportation
Okanagan District (1915), file 1752, section 1

Camp, another site that was more easily supplied was selected in mid-August near Edgewood, British Columbia, a small hamlet located on Lower Arrow Lake. Shortly after the site was chosen, 196 enemy alien prisoners were transferred from Vernon to the new Edgewood Camp, and there, with authorities ever mindful of the costs, they were tasked with cutting a road westward from Edgewood through the forest mountains, picking up where the unfinished road at Monashee had left off, and pushing it on through to Vernon. When finally completed in August of 1916, the Vernon-Edgewood road was praised by the supervising engineer as an outstanding piece of work. British Columbia's minister of public works, Thomas Taylor, also personally inspected the construction. Travelling its entire length by car, he described it more prosaically as "money well spent."[7]

As for the original Monashee internees who had been sent to Mara Lake, they would cut the rock cliffs that had for the longest time posed an obstacle to creating a traversable road connecting the Okanagan with Revelstoke to the north-east and the mountain pass beyond. In the past, construction on the road had entailed heavy labour costs that had proved to be prohibitive, and the work was never properly undertaken. Consequently, when it was made available, the importance of internment labour for this particular project was not lost on the minister for public works, the Honourable Thomas Taylor. As the minister responsible for roads and transportation Taylor understood and appreciated the potential benefits that could accrue to a government that was hard pressed to find adequate funds for the roads that were so essential to the province's continuing growth and development and for which there was so much demand.[8] It

Top: Prisoners at work, Edgewood

Left: Rock work, Mara Lake

was a timely and unexpected opportunity, one that would have to be made good use of and could be made to work if expenses were kept to a minimum. Despite adverse conditions, the Mara Lake internees would have to work with hammer and chisel, carving out the imposing rock sections and extending the right-of-way through the rugged terrain. The final results, however, were far less than expected, since the internees proved to be uncooperative. Despite the use of corporal punishment and other measures that had little effect, the authorities were forced in May of 1916 to scale down the operation after relocating a majority of internees elsewhere. The Mara Lake Camp would eventually close in July 1917.

Another camp was set up at Morrissey, British Columbia, to initially house 164 enemy aliens who had been arrested at the Crow's Nest Pass Coal Company in the town of Fernie. The start of the war had witnessed a sharp rise in anti-immigrant sentiment. Across British Columbia, public calls were regularly issued for the wholesale internment of enemy and nonenemy aliens alike. The mood in Fernie was no exception. Native-born coal miners at the local colliery resented the employment of enemy aliens while "loyal" Canadians were fighting overseas, and during a period of particularly heightened tension, they refused to work until the enemy aliens were dismissed. When they were, local authorities invoked PC 1501 and arrested the lot. Since the facilities at Vernon were overflowing with prisoners, Internment Operations, acting on the initiative of the district Conservative association, leased property owned by the same colliery in the nearby deserted town of Morrissey in order to intern the detained aliens there.[9] The prisoners were sent the few kilometres down the road to Morrissey, where they occupied a three-storey hotel and worked

METHODS AND POLICY

Mr Bruhn, [road superintendent] who is in charge of the work at the internment camp at Mara Lake, was in a couple of days ago and stated that the work is not getting on nearly so satisfactorily as it did some months ago. The reason he indicated was that the interned men were advised by the military authorities that they did not have to work if so inclined. He gave, as an example, a specific instance which took place a few days ago when one of the prisoners refused to do as he was told. The Sgt Major ordered that he be placed in the 'Black Hole.' Lieutenant Day, however, who was present, addressed the Sergeant Major before the crowd saying "You know we can't make them work if they do not want to." The result of this method of handling is, as you may understand, most unsatisfactory. Furthermore, some time ago, an Order was issued that only half of the wages earned by them should be allowed for goods at the canteen. This is interpreted to mean that those who only work half-time get all the wages earned and those who work full-time only receive half or thereabout. As there is a prevailing opinion amongst them that they will not get anything that is held back, you can readily see that this Order is not conducive to their working continuously. From what I can learn, it is evident that there is very little satisfaction in working these men under the existing conditions.

J. Kilpatrick, Engineering Office, Revelstoke
to J.E Griffith, BC Deputy Minister Public Works
15 August 1916

Referring to your communication of the 15th instant regarding the internment camp at Mara Lake, I spoke about this to Colonel Wilson yesterday, but, as the Order came from Ottawa, he is helpless in the matter. He has advised them, however, that it was a mistake to interfere when everything was running smoothly. Regarding Lieutenant Day's remarks to the Sergeant Major, it was not good policy to say so before the crowd, but as a matter of fact he was correct – they are not military prisoners of war rather civilians.

J.E. Griffith, BC Deputy Minister Public Works
to J. Kilpatrick, Chief Engineer, Revelstoke
18 August 1916

BC Ministry of Transportation
North Okanagan and Salmon Arm Districts (1916), file 211

on the road that cut through the Purcell Mountain range. Ironically, scores of internees were eventually paroled to the same company when a labour shortage forced company management to petition for their release, and they resumed their former positions alongside the co-workers who had engineered their initial arrest.[10]

Perhaps the most innovative and important development in the West was the creation of sizable internment camps in the Dominion Parks of the Rocky Mountains. Faced with a much reduced budget as a result of the reallocation of resources to the growing demands of the war effort, the Department of Interior – the government agency responsible for the parks – readily accepted an offer to use internment labour for park development.[11] Internment Operations and the Dominion Parks Branch agreed to a division of responsibilities. The park service would provide suitable sites for the camps, the transport of supplies and foremen to regulate the work, and payment for the labour, while the guarding, control, and feeding of the aliens was to be the responsibility of the Internment Directorate. With this understanding, detention camps, financed by credits taken directly from war appropriations, were soon established in Canada's Rocky Mountain national parks. This arrangement would satisfy the minister of the interior, W.J. Roche, who stated that it was his preference to put the internees to work rather than "allowing them to eat their heads off."[12]

Not all the internees were treated alike. Among German and Austrian internees there was a clear social and class distinction. Enemy aliens of German nationality – "commercial agents and men in similar positions" – were considered to be unaccustomed to physical work and not easily given to mixing with working men. The "Austrians," an amalgam of individuals from the minority nationalities in the Austro-Hungarian Empire, were decidedly underclass and thought to be accustomed to hard labour as working men. This distinction had some practical and political implications for the Directorate of Internment Operations. The latter group would work for their keep. The former, on the other hand, would have to be treated differently, especially since the German imperial government had a continuing interest in their co-nationals abroad. Furthermore, as a literate and relatively informed group, the Germans, it was thought, would collectively protest their ill treatment. Critically for Canadian authorities, the Hague regulations made an allowance for first- and second-class prisoners of war; that is, they distinguished between "officers" and enlisted men. The difference between the two groups could therefore be maintained more or less in practice by sending second-class enemy aliens to work in the Canadian bush. As a result, the "Austrians" – Ukrainians and others – were transferred to the hinterland of the Canadian Rockies, while the majority of Germans were detained in the urban centres of Amherst, Halifax, Vernon, and Fort Henry in Kingston.[13]

The first camp in the Dominion parks, which was created on 14 July 1915, immediately after General Otter offered prisoner labour for park development, was a tent compound at the foot of Castle Mountain, a disarmingly beautiful setting that had a hypnotic effect on all those who came to visit.[14] Situated at the terminal point of the still-unfinished Banff–Lake Louise Coach Road, the camp remained operational for two years while 660 men forced a right of way through the thick forests and dense underbrush. After a less than auspicious beginning, Castle eventually became an efficiently run facility, the camp administration taking

Work on Banff–Lake Louise Road, Castle Mountain

steps to ensure that attempts at escape would be at a minimum and discipline among prisoners strictly enforced. From the perspective of park officials, it was in this regard a successful camp. There were few interruptions in the schedule of work, and the labour force was, if nothing else, dependable. These considerations were critical, because perhaps no project was more important to parks officials than the road to Lake Louise, since it would eventually attract even more traffic and tourists to the area.

Tourism was seen as the key to the future of the Dominion parks, but the European war, which was diverting resources and funds, threatened to set back years of work and planning in park development. Consequently, when General Otter offered the use of internment labour, the parks commissioner, J.B. Harkin, not only gladly accepted but sought to capitalize on the opportunity by establishing more camps elsewhere. It was Harkin, for instance, who proposed to Otter that in addition to the Castle Mountain Camp another station for two hundred civilian prisoners should be created at the foot of Mount Rundle; a recommendation he would later change in favour of creating a single larger camp at the historic Cave and Basin site on the outskirts of Banff, where prisoners and soldiers alike from the Castle Camp would retreat before the onset of each winter.

"Such Conduct Cannot Be Tolerated"

I am in receipt of a communication this day from the General Officer Commanding Internment Operations in which he states that:

> The various complaints made to you by prisoners as to the rough conduct of the guards I fear is not altogether without reason, a fact much to be regretted, and, I am sorry to say, by no means an uncommon occurrence at other stations.

In this connection it is hoped that you will do your utmost to warn the guards that such conduct cannot be tolerated.

Gen. E.A. Cruikshank, Commanding Officer, Military District No. 13
to Captain P.M. Spence, Commanding Officer, Banff Internment Camp
20 December 1915

National Archives of Canada
Record Group 24, vol. 4721, file 1

Harkin's ambitions, however, did not stop there. Viewing the use of internment labour as a potential bonanza, he proposed an additional internment camp for three hundred enemy aliens at Jasper and another for two hundred to be situated at the junction of the Ottertail and Kicking Horse Rivers, near the village of Field in British Columbia's Yoho National Park.[15] Harkin's desire to retain the services of as many of the internees as possible was unimpeded by his knowledge and awareness that the civilian prisoners were, in his own words, "entitled to certain consideration." But the push to engage in as much of the work as possible overrode any misgivings or reservations Harkin may have had.

The two hundred enemy aliens who were eventually assigned to Yoho originally came from a camp created earlier in Revelstoke National Park. The municipality of Revelstoke, believing that the local economy would benefit by opening up the road to the summit of nearby Mount Revelstoke, had petitioned Internment Operations to establish a camp in the park.[16] Supported by senior park officials, who believed that internee labour could be used to good effect to complete the project in short order, the request was approved by the Internment Directorate in July 1915. Within a few short weeks, two hundred enemy aliens were sent down from the holding stations in Vernon and Brandon to construct log buildings and begin work on the road. It soon became clear, however, that the decision had been made in haste, without due consideration for the difficult nature of the site. The water supply was insufficient, and the high altitude of what was now called Revelstoke Camp would not only make quartering through the winter impossible but also affect the health of a number of the prisoners. When a prisoner strike brought on by "extreme

"Site in Park Selected"

The establishment of a camp for interned aliens in Revelstoke Park was arranged for on Thursday by J.B. Harkin, Commissioner of Dominion Parks, who in company with W.W. Cory, Deputy Minister of the Interior, arrived in the city on Wednesday and left for the East on Thursday night. On Thursday, Mr Harkin and Mr Cory drove up the automobile road and selected a site for the internment camp. The spot chosen is where the present road camp stands, about eight miles up the automobile road. Mr Harkin afterwards telegraphed General Otter saying that he had selected a provisional site and asked that the military authorities in charge of interned aliens inspect it. As soon as the approval of the military authorities is received work will be started on preparing the camp for the reception of the aliens.

The number of aliens interned in the camp will be 225. This is the greatest number that could be economically employed on the road. Probably 100 guards will be employed in watching the camp. It is expected that the guards will be members of the militia, provided under the instruction of the District Officer Commanding. The campsite will be cleared and fenced with barbed wire. Inside the enclosure will be erected two log sleeping cabins, a mess house, cookhouse and hospital. Immediately outside the wire fence, the land will be cleared for the patrol of sentries and buildings will be erected for the accommodation of the guards. It is expected that the internment camp will be in operation in two weeks time and it will remain as long as it is possible for road work on the mountain to be continued. It is expected that the road to the summit will be completed before snow prevents further work.

War prisoners are not criminals, says Mr Harkin, but are in many cases merely citizens of countries with which the Empire is at war who happened to be in Canada at the time that hostilities were declared. Under international law they may not be treated as ordinary prisoners but are entitled to certain consideration.

Mr Cory, the Deputy Minister, who accompanied Mr Harkin, is spending a holiday in the mountains. Both Mr Harkin and Mr Cory were delighted with the view from the automobile road. "It is unique," said Mr Harkin, "and impresses everyone who sees it, and if it delights even those who are accustomed to the mountains, how must it strike visitors unused to the beautiful mountain scenery. Once the road is completed the park cannot fail to draw tourists. When the road is finished there will be work in the park for 25 to 50 years in building roads and bridle paths and in other development work."

Revelstoke Mail Herald
Revelstoke, British Columbia
31 July 1915

cases of soldiering" broke out, Otter, who had been sceptical from the outset about the viability of the camp, ordered it to be dismantled. In total, the internment operation at Revelstoke lasted slightly more than one month (from 6 September to 19 October), and although municipal officials tried desperately to secure internees the following summer, in order to complete the unfinished motor road, the camp was never reopened.[17]

With the closure of Revelstoke, the prisoners were transferred to Yoho National Park, where they began building the Otter Camp – named in honour of the director – on the open and exposed flats at the juncture of the Ottertail and Kicking Horse rivers. Commercial interests – the proximity of the camp to a stand of marketable timber – determined the precise location. The decision was unfortunate, however, since it ignored the

Prisoners at work, Otter

Refusing to Work

I beg to acknowledge the receipt of your letter of the 3rd instant with reference to a report which reached you stating that the aliens in the Yoho Ottertail Camp are now on strike.

In this connection, I beg to advise you that, about the time the Chief Superintendent was in Field, there was some trouble down at the camp. A number of the aliens, on being marched out, refused to work and had to be taken back to camp. The Officer Commanding went into the matter with them. It appears their principal grievance is that they cannot see why they should be interned while hundreds of other alien enemies are allowed not only to go free but to earn their living in the country. They also do not appear to be looking forward to another winter in a working camp.

The Officer Commanding, I understand, informed them that, if they would resume work right away, he would place their grievance before General Otter and would advise them later if any action was proposed. They have, therefore, all resumed work, but of course this may develop again later on.

E.N. Russell, Superintendent, Yoho and Glacier Parks
to J.B. Harkin, Parks Commissioner
9 August 1916

I beg to advise you that, on August 10th, all the aliens interned at the Otter Camp, Yoho Park, refused to work. I received a message over the phone on this date from the Officer Commanding who advised me that the men had gone out at the usual hour in the morning, but, on arriving at the point where their work lay, they had all refused to do any work. I personally went to the camp and was present at an interview between the Officer Commanding and nine delegates of the aliens. Their grievance appears to be the same as stated in my last letter on this subject, when they went on strike about two weeks ago. At that time, the Officer Commanding promised the men that he would lay the matter before General Otter and advise them of his decision. General Otter's answer, I understand, arrived at the camp on August 9th, and was read to the men by the Officer Commanding, who advised them that no good could be done by striking.

Today, I have again seen the Commandant who informs me that the only aliens working are those cutting wood for themselves, or a few others engaged in some occupation in which they are personally interested. Those who refuse to work are being disciplined. He appears to have some hope that they may again go to work next week.

E.N. Russell, Superintendent, Yoho and Glacier Parks
to J.B. Harkin, Parks Commissioner
12 August 1916

National Archives of Canada
Record Group 84, vol. 124, file Y176

potential for flooding and the fierce prevailing winds that cut through the mountain pass, making the area virtually uninhabitable during the winter. Having spent one brutal winter there and fearing another, the disheartened prisoners began to resist the work schedule. And although the camp was relocated down the Kicking Horse to a more sheltered location in the vicinity of Boulder Creek, the inmates remained anxious and finally began a complete work stoppage when it was learned through uncensored correspondence that internees at other camps were also refusing to work. Despite individual punishment and threats, when parks officials urged the military to compel the internees to work, the strike at the new site held, forcing camp authorities to concede that "unless more strenuous measures were resorted to" work could not be expected to resume.[18]

In the end the threat of a prison riot and increasingly bold attempts at escape convinced officials that there was no further point in keeping the camp open. In late October 1916 the Otter closed. But the authorities were reluctant to release this militant group of internees into the unaffected prisoner population of the nearby camps, and the prisoners from Yoho National Park were sent to the isolated Spirit Lake Camp in the Abitibi region of northern Quebec.

The Jasper camp proved less troublesome but was not without its own problems. The internees at Jasper – two hundred enemy aliens transferred from Brandon – were tasked with developing the network of roads in and around Athabasca River, an area "destined to be one of the most popular health resorts in Jasper Park, as the waters of the lakes in this region [were] excellent for bathing and boating." It was an ambitious project, and the work proceeded as well as could be expected under the circumstances. But

"Interned Aliens Try to Tunnel Out of Stockade"

But for a timely discovery by guards at the Field [Otter] Internment Camp, some fifty odd prisoners would now be roaming the hills and possibly many would have made their escape on Thursday morning. The prisoners had tunnelled fourteen feet from the tent which they occupied toward liberty and it was only necessary to make but eight feet further to have brought their tunnel outside the radius of light from the stockade.

The tent is situated 12 feet from the stockade and occupied by fifty-three men, mostly Germans and Austrians, with a few Romanians and a Galician. Somehow they had obtained a shovel; with this and table knives they dug the tunnel, hauling the dirt out by a box covered with pieces of blanket to muffle the sound. The prisoners worked at night and stowed the dirt under their bunks and on the floor, trampling it smooth. When lined up and searched, each prisoner was armed with a dirk, made out of table knives with sharpened points and a saw edge on one side, ready to give battle with the guards when the getaway was made.

For some time the prisoners have been unruly and surly, having refused to work on the road with the exception of the Romanians, who wanted to work. Recently the Germans and Austrians refused to cut wood even for the camp, and as punishment were put on short rations and deprived of certain privileges they usually enjoyed.

Following the attempted break out, some of the prisoners attacked and badly mauled a Galician prisoner whom they suspected of having given information to the guards of the plans for escape. Then suspicion fell on one of the Romanians, and he became so afraid that he had to be allowed to sleep in the guardroom. As yet no trace has been found of the Austrian who escaped some weeks ago.

The Golden Star
Golden, British Columbia
5 October 1916

as with the other camps, there were hunger strikes, and when the work was especially dangerous, the internees refused to work or, after being disciplined, worked under protest, tending to slow down progress considerably. There were also numerous escapes that had a debilitating effect on the camp's operation, since parties of troops and local hired scouts had to be sent out in pursuit. Recapturing the prisoners proved both time consuming and expensive, prompting the suggestion that the costs incurred be deducted from their accounts. The great number of escapes, successful and otherwise, eventually led to a reassessment of the continuing viability of the Jasper Camp. When applications were made for the release of the internees from railways and coalmines in the late summer of 1916, the Jasper prisoners were systematically paroled into their custody.

The work regime at the camps in the national parks was varied. Although road construction was emphasized, internees were engaged in a range of activities, especially after it became known that they could be used to complete all manner of projects. At the Banff site, aside from quarrying and crushing stone, the internees cleared land for the proposed St Julian subdivision. They repaired roads and sidewalks in the town, extended the Banff Springs golf course to eighteen holes, and constructed the clubhouse. At the Buffalo Paddocks they cleared bush, while at the Recreation Park they drained and filled land depressions. Elsewhere in and around Banff, they reclaimed land for tennis courts, a shooting range, and horse pasture, while improvements on the toboggan and ski run on Rundle Mountain were also undertaken. Nor were local requests ignored. An ice palace was built for the winter carnival, trails for the Alpine Club of Canada were cut, and at the urging of business people in nearby Lake

Of Golf Links and Oher Imperatives

I wired you on Saturday suggesting that you put as many aliens as possible on the work extending the [Banff Springs] golf links. As the aliens will not be at Banff a great while longer this season, and as it is desirable that all possible work on the golf links should be done by alien labour, I hope you will be able to arrange for a much larger gang than you have been operating there in the past.

You must keep in mind that our appropriations for the current year are very small. It is imperative that new work of this kind should be done as much as possible by the aliens. If we do not have it done by them, there will be very little chance of getting it done by day labour.

J.B. Harkin, Parks Commissioner
to S.J. Clarke, Parks Superintendent
8 May 1917

National Archives of Canada
Record Group 84, vol. 70, file R313

Enemy aliens at work on winter palace, Banff

Minnewanka, a temporary camp was constructed to clean the townsite and shore up the wharf used for cruise ships on the lake.

In Jasper, apart from road construction, the internees cut and prepared fence posts for the Buffalo Preserve at Elk Island National Park near Edmonton and built several bridges not only across the fast-flowing glacial waters of a number of mountain rivers but across treacherous gorges as well – dangerous work for those unaccustomed to such things. Other more mundane jobs also became part of their routine, including repairing water mains, collecting and raking loose stones, and generally making cosmetic repairs to roads in and around the town, to impress park visitors. When the Otter Camp was relocated to Boulder Creek, the internees were deployed in Yoho National Park around nearby picturesque Emerald Lake to prepare a network of trails for avid outdoor adventurers. However, it was at the initial campsite that they were used most effectively, the camp being situated to take advantage of a large stand of trees lost to a recent forest fire. The trees were to be cut down and shaped to produce railway ties and mining props and later to be sold for revenue. It proved to be a lucrative venture, the Internment Directorate showing a net balance of more than sixteen thousand dollars in its final accounts as a result of sales from this and other pulpwood.[19] Finally, as elsewhere, the internees in Yoho National Park were put to work on road construction, building the scenic Natural Bridge Road and surrounding fireguards.

Although the Revelstoke, Otter, and Jasper camps did not entirely fulfil all expectations, the work done was nevertheless considered "substantial" and "valuable."[20] Internment labour at Revelstoke, for example, allowed the park service to reduce its dependence and therefore its expenditure on day gang-labour by one third over the course of the entire project. It also allowed the Parks Branch to engage in construction on a job that admittedly could not otherwise have been undertaken during the war.[21]

Like the other work performed by the internees this work was tremendously important, in that it enabled the park service to maintain a significant level of activity at a critical time, the success of which was later leveraged in its bid to secure additional support and funding for further development. That the prisoners were civilian aliens of enemy nationality was acknowledged, but their status did not appear to have any bearing on their treatment. The irony of their deployment on a project that at the time was already being touted as having national significance was also not a factor. The key consideration was that internment labour was both profitable and useful. There was no greater evidence of this than at Banff and Castle Mountain, where the new and improved road, perfectly graded, impressed not only the visitors, whose traffic increased markedly as a result, but the officials in Ottawa, as well, when they came to inspect. So taken was the commissioner of the parks with the results of internment labour that he hoped that gaol and penitentiary labour would be secured after the war for similar work in the parks.[22]

But perhaps the most telling indicator was the indelible impression that was left with a young soldier John Anderson-Wilson, who participated in the operation. Many years later, having witnessed first-hand the results of the internment labour, he still spoke of the "marvels" that had been accomplished by the internees, noting in his remarks the principle upon which the operation was based: "Anybody who asked [us] to do anything,

New grade, Edgewood

we provided the slaves."[23] It was a perspicacious statement, but not necessarily because it was true. On the contrary, the description was clearly overdrawn. Rather, it captured a notion often associated with human conflict, an idea deeply imbedded in the crusading spirit of war: that in war there are the victors and the vanquished and to the former go the spoils. It is a proposition, however, that makes sense only if the defeated exist outside the body politic. But the problem, of course, was that the internees, by reason of their settlement, were very much a part of the country called Canada.

6

War, Patriotism, and Internment: The Debate over Otherness

BY EARLY 1916, THE LABOUR SUPPLY in Canada had diminished considerably as the ongoing carnage in Europe consumed vast numbers of men and the economy began to grow. As a result, more generous wages were being offered by employers to attract workers to their ranks. Applications for the release of interned enemy aliens were also made to the Internment Directorate as a way to acquire much-needed labour. Consequently, from the middle of 1916 to the middle of 1917 large numbers of internees were released and paroled to companies that could guarantee their employment.[1] The exodus of prisoners from the camps during this period led to the closure of some sites and the consolidation of others. The Jasper Internment Camp ceased operations in August 1916, followed by the Edgewood and Otter Camps in September of the same year. Operations at Banff and Mara Lake were scaled down and eventually ended by July 1917. Among the frontier camps in the Rocky Mountain region, only Morrissey continued to operate after the summer of 1917, being dismantled just weeks before the armistice was signed.

Troop barracks and prison compound, Morrissey

Like several other camps remaining across the country, Morrissey served as a place of detention for the more truculent prisoners. It was felt that they could not be released because of their perceived hostility to Canadian authority, a not altogether incorrect assessment, for many had resented their internment and were less than forgiving in their views. Some effort was made to isolate the more troublesome characters, sending them to either Vernon, Amherst, or Kapuskasing, where they were held until deportation. That measure proved successful, enabling Morrissey, at least, to function as a working camp throughout 1918, although the labour by this time was voluntary. When Morrissey closed on 21 October 1918 – effectively marking the end of internment operations in the Canadian

Rockies – the internees who had not yet been paroled were either sent to Vernon, the last holding facility in Western Canada, or as in the case of fifty-five who had volunteered, relocated to Munson, Alberta, and later Eaton, Saskatchewan, where they worked for several more months laying the CNR rail line before finally being shipped off to Amherst, Nova Scotia.[2]

Kaisers in effigy, Banff

The militancy of the interned enemy aliens, which had increased considerably throughout 1918, did not go unnoticed in official circles. The remaining internees were characterized as especially intransigent, raising a philosophical question: what precisely was Canada's responsibility with respect to the internees? Of course, this question also touched on the subject of Canada's obligation to enemy aliens more generally, but it could perhaps be addressed neither fairly nor calmly when virtually all aspects of national life were being filtered through the lens of war and the sacrifices being made.

Indeed, in 1918 as the tally of casualties mounted and the prospect of an endless war seemed real, the less measured statements made within the wider public began also to colour the rhetoric of politicians who echoed those sentiments. In the current crisis, there was a sense that Canada faced an abyss in which there could be no allowance for weakness of spirit or generosity of the heart. To give in to those impulses was to entertain peril. Total war had raised the ante, and no quarter was to be given the enemy. Hence, in a parliamentary debate on a motion concerning compulsory alien labour that included a discussion of internment, it was argued, for instance, that liberal sentiment had made a mockery of the policy, for it was a cruel fact that as patriotic Canadians were being sent overseas defending liberty with pittance for pay, the interned enemy alien – given "the opportunity of working at 25 cents a day" but now refusing to work – was living "on the fat of the land."[3] Better fed than ordinary working men, the internees, it was suggested, were living in comfort, an unconscionable situation requiring the immediate action of the government, which should "see that these men [were] treated in a very different man-

The Opinion of the Great War Veterans

May it please you and give consideration to the humble petition of this gathering of representatives of the Great War Veterans' Association of Canada which follows:

WHEREAS there are in Canada a great number of people of alien origin,

THEREFORE BE IT RESOLVED that it is our opinion that the aliens of enemy origin in our midst should be employed in work of national importance, or in industries essential to the winning of the war, under proper surveillance, and their employer for the time being made responsible for them; and that their earnings over and above an amount equal to the pay and allowance of a Canadian soldier be taken by the Government for war purposes; or failing their being so employed, that such enemy aliens be interned.

FURTHER that measures be taken at once to make the Military Service Act applicable to allied aliens in the same manner and to the same extent as to the citizens of Canada, either by negotiating the necessary treaties or conventions with the remaining Allied Countries upon lines similar or the same as those provided for in the conventions recently adopted by the United States of America and Great Britain, or, failing the obtaining of such treaties or conventions that such allied aliens be forthwith given the option of enlisting voluntarily in the Canadian Forces or being deported to their country of origin as is being done by the Government of the United States under the Alien Slacker Bill just passed by an overwhelming majority in the House of Representatives.

We respectfully beg to submit the following in addition to the foregoing:

No.1 That no enemy alien shall any longer hold public office and that all questions having to do with the alien be taken out of the hands of the provincial authorities and undertaken altogether under direct federal supervision.

No.2 That the Canadian government establish an Alien Registration Bureau, the same as obtains in Great Britain, in which every Neutral, Allied, and Enemy Alien shall be registered so as to be used to the best advantage in the National Service and, moreover, that all aliens be compelled to wear a badge or token, prominently displayed, designating that he is an alien in his class.

No.3 That all enemy alien newspapers or periodicals should be suppressed, or, failing, it should then be insisted that all editorials be printed in English.

(cont.)

No.4 That no person of alien birth, whether naturalized or not, shall be permitted to have in their possession any firearms of any description.

No.5 WHEREAS greater production of food is of vital importance;

AND WHEREAS especially in the Western Provinces the farmers are labouring under a great disadvantage owing to the unreliable conditions of alien labour;

AND WHEREAS much hardship has resulted from breeches of contract [by alien labour] during harvest time;

THEREFORE BE IT RESOLVED that it is the opinion of this conference of Great War Veterans of Canada, assembled, that a law be passed punishing with a heavy fine such breaking of contracts where it is shown that the employer has fulfilled his contract.

Petition, Great War Veterans' Association, Ottawa
to the Rt Hon. Sir R. Borden, Prime Minster of Canada
28 March 1918

National Archives of Canada
Manuscript Group 26, vol. 241, reel C-4415

A stubborn prisoner, Castle Mountain

ner," especially in view of the rumoured harsh treatment of Canadians languishing behind German barbed wire. The interned enemy alien needed to be compelled to work for Canada, irrespective of international law or the protests of enemy governments, for as W.F. MacLean, member of parliament for South York, rhetorically put it, "What was the good of having a moral discussion with an unethical people like the Germans?" Firmness and resolve was what was required.

But if interned enemy aliens could be made to work for the national cause, then why not enemy aliens in general? Indeed, according to R.C. Cooper, the representative for South Vancouver, enemy aliens were enjoying the benefits of liberty but were contributing "absolutely nothing" to either the country or the war effort. How was it that those who cared not a whit for the country now stood to gain by it? What was or should be expected from them? The answer was no less than for all enemy aliens to be conscripted to work for the state, since as "natural road builders" they could build and repair the roadbeds necessary to meet the transportation needs of the country. "A fair wage" would, of course, be provided for the sustenance of both them and their families, "but the balance of their wages … should be taken and applied for the benefit of the state." Emphasizing the necessity and urgency of such action, another speaker suggested that there could be no compromise, for "the time has arrived, in the great stress of war, when international law might be forgotten, and when the Act of Confederation might be overlooked, if necessary, to meet the situation."

Encouraged, other members of parliament chimed in, offering their own views on the subject. Why, for instance, should the matter stop with the enemy alien? Could not alien labour of allied nationality also be conscripted for more productive purposes? Enjoying the same privileges as Canadian citizens but wearing their responsibility lightly, it was pointed out that Greeks, Swedes, and a host of others – engaged in nonproductive labour while avoiding service – were doing little to help win the war. Since they had failed to demonstrate sufficient patriotic sentiment, it was suggested that they should be sent back to their countries of origin, where at least they would be forced to serve.

As for the Chinese and Japanese, it was argued that they were unassimilable peoples who did not know what it would take to be a citizen.[4] Patriots alone would bring the war to a successful conclusion, and ordinary aliens who would not bear their share of the burden of war should be forced to do so. "These men within our gates enjoying all the privileges that the Canadian citizen enjoys, and making high wages as well," it was declared by the controversial R.F. Green (West Kootenay), "should be compelled to take up the duties of citizenship or else be told to leave this country, and until they do bear their full share of the burdens of this war, they should not be allowed to use the food produced by the other citizens of Canada."

Talk of patriotism invariably evoked comments on the situation in Quebec, where below-average enlistments were taken as evidence of a less than appropriate attitude toward the war. In keeping with the character of the discussion, the Tory member for Comox-Alberni, H.S. Clements, proposed that twenty thousand, if not possibly seventy-five thousand, French Canadians might be conscripted as "alien labour" and taken to British Columbia, where they would be placed "in their natural element,

"Put an End to this Ill Treatment"

On April 2nd 1918, we were sent from the Morrissey detention camp to the detention camp at Vernon, British Columbia. We can testify under oath that, since the middle of January 1917, German and Austrian civil prisoners of war at the Morrissey detention camp, at different times, have been cruelly treated by guards and especially by the camp police. Some fellow prisoners have been arrested under false charges and taken to the guardroom by the camp police. On the way to the guardroom and inside the guardroom, they were hit with fists and kicked by guards or camp police. Several of these prisoners fell sick as a consequence of such ill treatment, especially Nos. 335 and 188. No. 335 is at present in the isolation hospital in a hopeless condition. Both these prisoners with some others were put into close confinement in April 1917 because they refused to do work outside of the wire fence; work which was not in the interests of the inmates of the camp. While in the cells, they were treated in the most brutal manner. Other civilian prisoners of war, who refused to do such work, were often sentenced to close confinement too, and during their time in the cells were brutally treated and forced to do the most degrading kinds of work for the guards under the threat of bodily punishment for refusing to comply. Some prisoners, who called the attention of the Officer Commanding to these conditions, were told that they were liars. In spite of the fact that we have told this to the Swiss Consul, Mr S. Gintzburger, during his first and last visit to the camp during our stay there, August 24, 1917, and notwithstanding our vain efforts to keep in touch with the Consul, the treatment of the prisoners has not materially changed since that time. Only a few days before we left the camp, on March 28, 1918, prisoners were brutally treated by the camp police.

The civil prisoners at the Morrissey Camp are not allowed to lay down on their bunks during the daytime – even if they feel ill, tired, and hungry – without a permit from the Medical Sergeant, who gives his opinion whether the prisoner is sick or not. Prisoners, found on their beds at daytime, were punished with as much as 6 days in the cells at half rations, and, in spite of their feeling unwell, were forced to do humiliating work for the guards in the guardroom. Anybody refusing to do this was threatened with bodily punishment.

(cont.)

We have done everything possible to better the conditions of our poor German and Austrian fellow prisoners who are detained at the isolation barracks or at the hospital. Fresh milk, the chief nourishment for sick people, is not to be had at these hospitals, and when we received $100 from Mrs Schroeder, Winnipeg, for the benefit of the camp, the prisoners resolved that this money should be used to buy milk and other proper food for the sick. The camp Medical Officer, when told of our intention, replied that no fresh milk was obtainable even if we paid for it ourselves.

In this matter we do not expect any help from our legal representatives, the Consuls of Switzerland or Sweden. During the week of February 24, 1918, we requested the Swiss Consul, Mr S. Gintzburger, in 15 different letters, to visit the camp at least, but up to this time we have received no answer whatever. Therefore, we sincerely request the German Foreign Office to cause immediately the necessary measures to be taken by the Imperial German Government in order to put an end to this ill treatment. Furthermore, we beg the Imperial German Foreign Office to make these facts known to the Austro-Hungarian Government in order that similar steps may be taken.

We declare our willingness, voluntarily, to support these statements with an oath.

Petition Addressed to the Imperial German Foreign Office
1 May 1918

National Archives of Canada
Record Group 6 HI, vol. 763, file 4738

the lumberwoods" and, under military supervision, forced to work where they would learn the values of Canadian citizenship. The remark was viewed as a provocation and produced an uproar in the House; the outraged Liberal member of parliament for Maisoneuve, Rodolphe Lemieux, denounced the speaker's reference to French Canadians as aliens as a sign either of ignorance or of fanaticism. The apology that was extended but rejected could not mask the bitterness and deep divide that had widened with the stress of war.

The debate about alien labour was extraordinary, not only because it was free-ranging and uncharacteristically candid but also because it revealed the depth of confusion and emotion over the subject. It forced a response from the justice minister, the Honourable Charles Doherty, who felt compelled to state the government's position and lay bare some of the mistaken ideas regarding the status of the enemy alien.

Taking stock of the debate, Doherty noted that with respect to the internees much had been said about international law and what Canada could or could not do. But in fact, as Doherty communicated in a surprising admission, the interned enemy aliens were not prisoners of war and therefore could not be compelled to work. The distinction between enemy aliens and prisoners of war had to be maintained. There was, for instance, the ethical aspect to consider. Would Canada stoop to the inhumanity of the "Hun"? As he hoped to convey, Canadian policy was preeminently one of benevolence, providing from the outset refuge for the many indigent and unemployed. The state had a responsibility to these individuals, he claimed, having denied them the right to leave the country. Doherty agreed that if they were prisoners of war, the Hague convention would apply and Canada could compel them to work, but he impressed upon the House that this was not the case. He insisted that "[we] are dealing with civilian residents of this country," and as such, there was an ethical and legal obligation to treat the internees in accordance with the rule of law.

There was also, however, the practical question of possible retribution, a matter of particular concern to the minister because of the growing number of inquiries from London that had been prompted by German diplomatic protests alleging Canadian brutality. Acutely aware of the international implications, Doherty stressed that reprisals would not fall on Canadians alone but would also be inflicted on every British subject. Canada, the minister argued, was not an independent nation, and he called Parliament's attention to the policy of His Majesty's Government, which was clear on this point: work performed by interned civilians was

Hard labour, Edgewood

Of Unsatisfactory Accounts and Possible Reprisals

War Office concerned regarding report by Mr Gintzburger, Swiss Consul for British Columbia and Alberta with respect to Morrissey Camp. Stop. They are greatly afraid of German reprisals on Canadians unless allegations of Gintzburger are disposed of. Stop.

Cable from Dominion Office, London
to Prime Minister Borden
11 March 1918

I wish to direct your attention particularly to the Colonial Office despatch of the 27th ultimo, transmitting a copy of a report by the Swiss Consul at Vancouver concerning his visit of inspection to the internment camp at Morrissey at the end of August last, which is somewhat unsatisfactory. You will observe that Mr Long suggests that the German Government may possibly adopt reprisals. I think you should lose no time in ascertaining whether there is any foundation to the complaints voiced by the Swiss Consul.

E. Newcombe, Deputy Minister of Justice
to Gen. Wm Otter, Director, Internment Operations
25 March 1918

Answering Consul Gintzburger's report sent to the Swiss Legation, London, before the Director of Internment Operations had seen it, or had opportunity to present his side of case. It refers to many things long past and satisfactorily adjusted.

In early days few overzealous officers used mild forms of coercion to induce prisoners to work. But this and any other complaints in the report, having foundation in fact, have since been remedied.

Suggested cable
Major General Wm Otter
to E. Newcombe, Deputy Minister of Justice
n.d.

National Archives of Canada
Record Group 6 HI, Volume 765, file 5294

to be of a voluntary nature only. How would an otherwise Canadian policy sit with Britain? Appealing to the patriotic sentiments of the Canadian parliamentarians, Doherty asked whether they would endanger Canada by placing the country "outside the pale of law and in the company of Germany … or, would we rather stand within the pale, and so standing be in the company of the Mother Country?" This rhetorical question disguised a deeper concern. As the minister of justice apprised of confidential details relating to the operational aspects of internment and tasked with responding to German allegations of Canadian wrongdoing, he was aware of the problematic nature of Canada's internment policy.[5] The real political costs were now impossible to ignore.

Doherty was, of course, aware of the frustration that resulted from the perceived advantages enjoyed by enemy aliens. Indeed, he felt that they should make a tangible contribution to the country. But compelling them to work was an altogether different matter. As a practical issue, he asked, how could this policy be implemented without resorting to the threat of wholesale internment? In the abstract, conscripting enemy alien labour might sound satisfactory, but it could not be carried out effectively. Moreover, very little productive labour, he explained, could in fact be obtained without employing force. Indeed, speaking with knowledge of the mixed results of the internment experience, Doherty questioned whether the parliamentarians intended "[to] send a slave driver out with the alien armed with a whip to keep him working, [because] short of doing that what course could you pursue?" Would it not have been better for enemy aliens to work freely when there was a great demand for labour, even though they received large wages? For Doherty, despite calls from his parliamentary colleagues to the contrary, internment was no longer a viable option.

Lost in the disquiet of the debate were the measured remarks of a lone voice, Fred Langdon Davis, Liberal member of parliament for Neepawa. Davis expressed his frustration and consternation with the concept of enemy alien. How was it that these immigrants had become enemies in the first place? He insisted that the concept was much too facile and that there could be no reasonable quarrel with the fact that it had "not a truth in it." According to Davis, these individuals had expatriated themselves for the purpose of becoming Canadian citizens and "If we treat such men as men and brothers, it will make Canadians of them; if we treat them in any other fashion, we will make of them an alien element in Canada." He was convinced that individuals of foreign descent were reliable and loyal and instinctively knew of their responsibilities, noting with pride the great number of "enemy" aliens in Manitoba who had offered to enlist. So why the animosity shown toward the alien by members of the House? Davis observed that lack of knowledge and familiarity with the immigrant, as well as insularity and smug self-sufficiency, had allowed for the placing "of a tag on a class of people … whom we do not very well know, and whom we are not in complete sympathy with."

Davis implored his colleagues to exercise restraint and good judgment and reject solutions that would harm the country and its future. Failure to do so, he cautioned, would lead inevitably to more of those situations where, for lack of sympathy and understanding, House members had felt justified in calling for the conscription as alien labour of French Canadians in Quebec. Could it be, he mused, that Canadians were so separated from each other that there could be no comity between Canada's peoples and that even now legislators were prepared to repeat the same mistake with respect to those who had recently come to Canada's shores? It was

Davis' hope and belief that this would not be the case.

In the end, the motion on compulsory alien labour, with all its talk about the further use of internment, was defeated – if only narrowly. It was perhaps a moot point, for the war would conclude but a few short months later. The armistice would be of little consolation, however, to the 1,964 internees who would still have to wait for more than a year in the three remaining camps – Kapuskasing, Vernon, and Amherst – until the terms of their release had been negotiated in a general treaty of the peace. German merchant marines who had been transferred and interned in Canada at the request of the Dominion Office numbered 840 of the total. The remaining 1,100 were civilian enemy aliens. The majority of this group were described as hostile to Canadian authority or as otherwise undesirable, the latter description used with reference to approximately 400 infirm or insane internees who were eventually deported. Among the

Road excavation, Monashee

group there were also, however, a few score who elected to return to their countries of origin, having had their fill of Canada.

As for the record, little was said and even less written about the experience, as much because of fear as because of the fact that most of the internees were unschooled and illiterate. Meanwhile, in the Canadian Rockies the abandoned camps, nestled in the wilderness were in a matter of years reclaimed by nature, leaving little trace of what had been there and what had once happened.

Enemy aliens, Morrissey

Notes

CHAPTER ONE

1 Paraphrasing Renan, Desmond Morton writes, "Nations ... are built from the experience of doing great things together. For Canadians, Vimy Ridge was a nation-building experience. For some, then and later, it symbolized the fact that the Great War was also Canada's war of independence even if it was fought at Britain's side against a common enemy." D. Morton, *A Military History of Canada* (Edmonton: Hurtig Publishers 1985), 145. The mythicized version of the Vimy Ridge experience and its symbolic national importance is discussed in Jonathan F. Vance's, *Death So Noble: Memory, Meaning, and the First World War* (Vancouver: University of British Columbia Press 1997), 223, 233.

2 Complicating the connection between militarism and national honour was the imperial obligation. And yet, as was often argued at the time, it was through the ennobling and civilizing experience of sacrifice and war that the

Dominion would earn for itself national honour and a reputation deserving of recognition. The role of the Great War and the nationalization of honour in the context of Canada's imperial tradition is briefly discussed by Geoffrey Best in his critical essay *Honour among Men and Nations: Transformations of an Idea* (Toronto: University of Toronto Press 1982), 51–3. See also Carl Berger, *Sense of Power: Studies in the Ideas of Canadian Imperialism, 1867–1914* (Toronto: University of Toronto Press 1973), 233–58.

3 In actual fact there were 8,579 prisoners of war in Canada, but of this number 817 were captured merchant marines transferred and detained in Canada for the duration of the war. See chapter 3, n3

4 See, for instance, R.H. Coats, "The Alien Enemy in Canada: Internment Operations," in *Canada at War: An Authoritative Account of the Military History of Canada from the Earliest Days to the Close of the War of the Nations*, vol. 2 (Toronto: United Publishers 1918–21), 161.

5 The imperial-nationalist conception of the Canadian character is deftly explored in Berger's *Sense of Power,* 128–52.

6 For a discussion of the effect of war on the interaction between dominant and minority cultures, see P. Panayi, "Dominant Societies and Minorities in the Two World Wars," in P. Panayi, ed., *Minorities in Wartime: National and Racial Groupings in Europe, North America and Australia during the Two World Wars* (Oxford: Berg Publishers 1993), 3–23.

7 Professor Waiser, writing more generally about the development of the national parks in the first half of the twentieth century and the role that internment and other prisoner labour played, remarks that "While regular Parks funding was being reduced in response to war and depression, several western national parks were more than compensated by the ready availabili-

ty of these large pools of men. And thanks to their muscle, the national parks experienced one of the most intensive development periods in their history. They tackled specific projects that had a clear purpose – they were not simply marking time." B. Waiser, *Park Prisoners: The Untold Story of Western Canada's National Parks, 1915–1946* (Saskatoon-Calgary: Fifth House Publishers 1995), 251.

8 Interestingly, J.B. Harkin, commissioner of dominion parks, emphasized that there was a natural connection between the development of the parks as a national symbol and the making of Canadians as determined and resolute citizens. It was, therefore, doubly ironic that alien labour should be used in the development of parks – would-be Canadians were labouring on a project that not only would be maintained for the benefit of "all the people of Canada" but would also make "better citizens" of Canadians. *Annual Report of the Department of the Interior, 1916* (Ottawa: King's Printer 1917) 6–9.

CHAPTER TWO

1 Canada, House of Commons, *Hansard* (Special Session of Parliament), 19 August 1914, 20.

2 On 27 July 1914 the Ruthenian Greco-Catholic hierarch in Canada, Bishop N. Budka, issued a pastoral letter urging the faithful recent migrants from the Austro-Hungarian crownland of Galicia, to help "our old Fatherland" and "brethern" from what was described as the "heavy hand of the Muscovite despot." The bishop was oblivious to the broader possibilities in the looming conflict, and as a result the message proved embarrassing, since Canada soon found itself at war with Austria-Hungary. The letter was

promptly retracted and followed by a missive that awkwardly, if not crudely, called upon the faithful to serve their "new fatherland." On the question of community politics and clashing personalities and ambitions, see F. Swyripa, "The Ukrainian Image: Loyal Citizen or Disloyal Alien," in F. Swyripa and J.H. Thompson, eds., *Loyalties in Conflict: Ukrainians in Canada during the Great War* (Edmonton: CIUS Press 1983), 47–9; and Orest Martynowych, *Ukrainians in Canada: The Formative Period, 1891–1924* (Edmonton: CIUS Press 1991), 315–23.

3 As it relates to the Bishop Budka affair, this point is expertly laid out by Stella Hryniuk in "The Bishop Budka Controversy: A New Perspective," *Canadian Slavonic Papers* 23, 2 (June 1981): 154–65.

4 See, for example, various RCMP reports in the National Archives of Canada, RG 18, vol. 1779, file 170.

5 The apprehension of those who feared the immigrant impress on the Canadian character is described by Berger, *Sense of Power*, 147–52. For a discussion of the role of war in shaping a collective identity, see Anthony D. Smith, "War and Ethnicity: The Role of War in the Formation, Self-Images and Cohesion of Ethnic Communities," *Ethnic and Racial Studies* 4 (1981): 375–97.

6 In his study of vagrancy and the law in turn of the century Calgary, David Bright comments that "Canada's nineteenth century middle class viewed vagrants and tramps not so much as a physical threat to orderly urban existence, but as a challenge to the implicit beliefs that underpinned that existence." D. Bright, "Loafers Are Not Going to Subsist upon Public Credulence: Vagrancy and the Law in Calgary, 1900–1914," *Labour/Le Travail* 36 (fall 1995): 41.

7 See D. Avery, *"Dangerous Foreigners": European Immigrant Workers and Labour Radicalism in Canada* (Toronto: McClelland and Stewart 1979), 65–8; and D. Avery, *Reluctant Hosts: Canada's Response to Immigrant Workers, 1896–1994* (Toronto: McClelland and Stewart 1995), 71–2.

8 Richard Speed, *Prisoners, Diplomats, and the Great War: A Study in the Diplomacy of Captivity* (New York: Greenwood Press 1990), 165–6. Some critics, viewing the security measures as an assault on American civil liberties, have described Wilson's actions as reactionary and heavy-handed. Given the administrative excesses of the United States Department of Justice and Postmaster General (Censorship), this criticism may be valid. But the continuing role of the courts meant that in practice the measures would be applied "judicially" and reservedly. In this regard, of the approximately 800,000 un-naturalized aliens of enemy nationality resident in the United States in 1917, some 6,000 cases were brought before the courts, leading to the internment of 2,300 individuals. This figure, comparatively speaking, is remarkably small. For a discussion of the attitude of the U.S. president and the policies of his administration, see Jörg Nagler, "Enemy Aliens in the USA, 1914–18," in P. Panayi, ed., *Minorities in Wartime: National and Racial Groupings in Europe, North America and Australia during the Two World Wars* (Oxford: Berg Publishers 1993), 191–215. For a more critical assessment, see P. Murphy, *World War I and the Origin of Civil Liberties in the United States* (New York: Norton 1979).

9 For an account of the labour situation in Edmonton and Calgary, see David Shultze, "The Industrial Workers of the World and the Unemployed in Edmonton and Calgary in the Depression of 1913–1915," *Labour/Le Travail* 25 (spring 1990): 47–75.

10 Public Record Office, Colonial Office Papers, CO 616/12/481-3, 26 October 1914. See also Martynowych, *Ukrainians in Canada*, 325–6.

11 David Saunders argues that British interests were not motivated entirely by security concerns. Rather, there was also the underlying fear that aliens repatriated to Europe via Britain would be held up en route and that the United Kingdom would become the de facto resting place for undesirables. See D. Saunders, "Aliens in Britain and the Empire during the First World War," in Swyripa and Thompson, *Loyalties in Conflict*, 110–11.

12 To offset the problem of unemployment among the Canadian-born, the Borden government urged and campaigned for the entry of unemployed Canadians into the ranks of the Canadian Expeditionary Force. This option was eagerly taken up by thousands of young men who foresaw little change in their economic prospects.

13 P. Melnycky, "The Internment of Ukrainians in Canada," in Swyripa and Thompson, *Loyalties in Conflict*, 198.

CHAPTER THREE

1 For a full account of the career of one of Canada's distinguished military officers, see Desmond Morton, *The Canadian General: Sir William Otter* (Toronto: Hakkert 1974).

2 The internment experience at Spirit Lake is described in P. Melnycky, "Badly Treated in Every Way: The Internment of Ukrainians in Quebec during the First World War," in M. Diakowsky, ed., *The Ukrainian Experience in Quebec* (Toronto: Basilian Press 1994), 51–78.

3 Eight hundred and seventeen were received from the British possessions of

Jamaica, Barbados, Bermuda, St Lucia, and British Guiana. A few were also received from Newfoundland. See G.P. Bassler, "The Enemy Alien Experience in Newfoundland, 1914–1918," *Canadian Ethnic Studies* 20, no. 3 (1988): 42–62.

4 Just as it is possible to overestimate the direct, personal effect of internment on the enemy alien population in Canada, which numbered some 120,000, it is also easy to underestimate the psychological impact of internment on the same population. As a percentage of the enemy alien population in Canada, the number of interned is, relatively speaking, small, certainly in relation to the figures in Europe. This observation, however, misses the general purpose behind the policies and security measures adopted, namely, the social control of the enemy alien population. Registration and reporting was used to monitor all enemy aliens who resided in urban or rural municipalities, but in highlighting the politically tenuous nature of their situation in the country, it was the prospect of internment that, effectively ensured their control. Local enforcement officials, for instance, often commented on the "sobering effect" of internment on the "foreign-born" population, and ethnic labour in particular, making them more pliant and complacent. This was in part the result of the seemingly arbitrary way in which the measures were being applied. Because it was arbitrary, internment, although not widespread, was still an effective, if blunt, instrument in ensuring social compliance and political calm.

In this regard, it is useful to note the difference between the American and the Canadian internment experience. The dissimilar role that internment would play in each country – one focused more narrowly on defined security threats and the other on social control – would translate, not surprisingly, into remarkably divergent numbers of internees. Whereas in Canada there were 7,700 civilian enemy aliens interned out of a general population of 120,000, in the United States 2,300 civilian enemy aliens were interned as part of a much larger total – an estimated 800,000 individuals who were classified as enemy aliens at the time of America's entry into the war.

5 Commenting on the episode, Otter claimed that "Some municipalities are attempting to take advantage of the situation to relieve themselves of the taxation necessary for the relief of the unemployed or destitute foreigners and I think that Port Arthur and Fort William are in this class." He would repeat this observation in his final report on internment. Morton, *The Canadian General*, 334; and Sir William Otter, *Internment Operations, 1914–20* (Ottawa: King's Printer 1921), 6.

6 As quoted in Morton, *The Canadian General*, 337.

7 M. Minenko, "Without Just Cause: Canada's First Internment Operations," in L.Y. Luciuk and S. Hryniuk, eds., *Canada's Ukrainians: Negotiating an Identity* (Toronto: University of Toronto Press 1991), 294.

8 The incident was reported in detail in the *Fernie Free Press*, 11, 18, and 25 June. The development at the Crow's Nest Pass Coal Mine came to the attention of the editors of the authoritative industry monthly, the *Canadian Mining Journal*. On the opening pages, copy was given to addressing the situation, albeit cautiously, with a view to accommodation. Critically, the main points of the issue were identified and articulated in a balanced and responsible manner:

> It is evident that some of the alien enemies in the mining camps are thoroughly in sympathy with the German government. Of these, some have

been foolish enough to give joyous voice to the successful accomplishment of criminal acts such as the sinking of the *Lusitania*. Avowed enemies such as these should be promptly interned.

On the other hand there are many Germans who are neither in sympathy with the German warlords nor who would conduct themselves in a manner hostile to the country in which they live. Such men we should not hastily throw out of work, for their labour is useful to the country as well as to themselves. Why should we make public charges of men who would otherwise be productive workers? And even if there were no economic loss, is it fair that these men, who are striving to live as becomes decent citizens, should be made to suffer because others of the same nationality have been unwise enough to approve openly of the mad policy of the Kaiser and his brood? The men employed in the mines have won their positions by their work, and, so long as their work is satisfactory to their employers and their conduct satisfactory to the public, it will be grossly unfair for anyone seeking personal interest to ask that they be refused employment. Editorial, *Canadian Mining Journal* 36, no. 6 (June 1915)

9 The rights of resident aliens, in this case, were legally protected interests that grew out of public sentiment and were granted by a national legislature. International law also expected equality of treatment but further guaranteed a minimum standard – basic human rights or essential freedoms – that could not be violated, a principle that was directly addressed and clearly articulated in the Calvo Doctrine. See Charles Fenwick, *International Law* (New York: Appleton-Century-Crofts, 1952); and G.H. Lloyd, "Nationality and Domicile with Special Reference to Early Notions on

the Subject," *Transactions of the Grotius Society: Problems of Peace and War* 10 (1925).

10 When the United States became party to the conflict, the inclination of the Wilson administration was to intern enemy aliens who wished to depart. On learning that such action would violate article 23 of the American-Prussian Treaty of 1828, which allowed nine months in which aliens could leave the country if they so desired, Wilson declared that the provision would be honoured on the grounds that "it would be unfortunate to open the war by tearing up a treaty." Richard Speed, *Prisoners, Diplomats and the Great War: A Study in the Diplomacy of Captivity* (New York: Greenwood Press 1990), 159.

11 See Minenko, "Without Just Cause," in Luciuk and Hryniuk, *Canada's Ukrainians*, 288–303.

12 See E. Satow, "The Treatment of Enemy Aliens," *Transactions of the Grotius Society: Problems of the War* 2 (1916); and J. Garner, "Treatment of Enemy Aliens: Measures in Respect to Personal Liberty," *American Journal of International Law* 12, no. 1 (1918).

13 Clause 5 of order in council 2721, 28 October 1914, declared that "No alien of enemy nationality shall be permitted to leave the country without an exeat from a registrar." Although clause 6 of the same order provided for the possibility of registrars issuing exeats if they were satisfied that certain conditions had been met, because it was a discretionary matter and because the overarching principle was not to permit enemy aliens to leave Canada, the provision under the clause was rarely met. Interestingly, that there was an obligation of sorts on Canada to have issued exeats was implied in remarks made by the Honourable C.J. Doherty, the justice minister, at the conclu-

sion of the war. See Canada, House of Commons, *Hansard*, 22 April 1918, 1018.

14 Significantly, Britain distinguished between "friendly" and "enemy" aliens; that is to say, "friendly" aliens were individuals whose origins could be traced to states that were at war with the Entente but whose own national aspirations, if anything, predisposed them to be more sympathetic to the Allied cause. The stated policy of Britain was "to afford [friendly aliens] them as favourable treatment as is possible in the circumstances." Among the Austro-Hungarian nationalities considered friendly were Czechs, Croats, Italians from Trieste and Trentino, Poles, Roumanians, Serbs, Slovaks, Slovenes, and Ruthenians or Ukrainians. The Armenians and Christian Syrians under Ottoman rule and Alsatians in the case of Germany were also classified as friendly aliens. NA, RG 13 B8, vol. 1368, file "January-February 1915," Secretary of State for the Colonies to the Governor General (copies to the Prime Minister, Justice and General Otter), 1 February 1915; and Sir Cecil Spring Rice, British Embassy, Washington to the Governor General (copies to the Prime Minister, Justice and General Otter), 22 February 1915. The issue of which nationalities constituted friendly aliens is explored in D. Saunders, "Aliens in Britain and the Empire during the First World War," in F. Swyripa and J.H. Thompson, eds., *Loyalties in Conflict: Ukrainians in Canada during the Great War* (Edmonton: CIUS Press, 1983), 99–124.

15 International law did provide for the internment as military prisoners of war of individuals who were considered to be "agents" attached to the army or of individuals, such as reservists, who could still provide service during the war. But even so, international law also made clear that certain conditions had to be met before they could be classified as military prisoners.

16 The Wartime Elections Bill was introduced by the Borden government in September 1917; it disfranchised immigrants who had been naturalized since 1902, because of their former enemy nationality. It was feared that without restrictive legislation the government of Sir Robert Borden would lose the wartime election.

17 *Hansard*, 10 September 1917, 5889.

CHAPTER FOUR

1 See J. Garner, "Treatment of Enemy Aliens: Measures in Respect to Personal Liberty," *American Journal of International Law* 12, no. 1 (1918).

2 See Richard B. Speed, *Prisoners, Diplomats and the Great War: A Study in the Diplomacy of Captivity* (New York: Greenwood Press 1990), 153–66; and Garner, "Treatment of Enemy Aliens," 55.

3 National Archives of Canada (hereafter NA) RG 24, vol. 4280, file 34-13-4; and NA RG 24, vol. 4513, file 17-2-40 (part 3) also contain copies of many other orders that added to or amended the original regulations for the purpose of guiding military conduct at the internment camps and stations.

4 See Desmond Morton, *The Canadian General: Sir William Otter* (Toronto: Hakkert 1974), 344; and D. Morton, "Sir William Otter and Internment Operations in Canada during the First World War," *Canadian Historical Review* 55, no. 1 (1974): 50–1.

5 Morton, "Sir William Otter and Internment Operations in Canada during the First World War," 58.

6 In response to a question in Parliament about the nature of the work performed by the internees, the Honourable W.J. Roche, minister of interior,

noted that it "was compulsory and not voluntary," explaining that this was the only way by which "we could get a lot of work done." When the legality of the activity was questioned, Roche replied that he expected no international repercussions, in light of the treatment of British prisoners of war in Germany. Significantly, by suggesting equality of treatment, Roche also implied that interned civilian enemy aliens and military prisoners of war were equal in status. In effect, he made no distinction between the two. Hence, there could be no political fallout. Since the Hague covenant provided for situations where prisoners of war could be made to work, Canada was simply following the same kinds of practices found elsewhere as they applied to the treatment of prisoners of war.

Roche was correct in surmising that there would be little in the way of international repercussions, but not for the reasons stated. Interned civilian enemy aliens, with the exception of reservists, were not military prisoners of war. This was well understood by authorities in London, who, having received a copy of a U.S. Department of State report on conditions at the Banff Camp, observed that the internees in Canada were forcibly engaged in work, in clear violation of the international code of conduct. A British Foreign Office official, however, noted that since the internees appeared not to complain "I do not think we should say anything about it." Canada, House of Commons, *Hansard*, 15 February 1916, 850; and UK Public Record Office, Foreign Office Papers, FO 383/239, "Prisoners of War Camp at Banff," 29 June 1916.

7 The first tentative steps at distinguishing between soldiers and civilians, combatants and noncombatants, occurred in the aftermath of the Thirty Years War, a grisly conflict that had decimated the population of East Cen-

tral Europe and threatened European recovery. It was felt that limitations had to be placed on warfare, and the foundation of a much later war convention was the result. The practice, for instance, of wholesale massacres, pillage, and laying siege to towns was generally accepted as being outside "civilized warfare." Significantly, the approbation granted to "humanizing" war would allow for certain protections to be extended to civilians. Of course they were not always adhered to, as is evidenced by Napoleon's effort in 1809 to suppress the rebellion on the Iberian peninsula through the use of terror or by the unspeakable acts committed by Grant and Sherman against the citizenry of Vicksburg and Atlanta during the American Civil War. Nevertheless, notwithstanding these and a few other exceptions, the principle of extending protections to civilians was generally observed, at least until the industrial wars of the twentieth century. Critically, however, although a distinction was made between solider and civilian, it was still to be determined – beyond the simple recognition that noncombatants were to be extended the courtesy of noninterference – what specific rights the noncombatant was entitled to as a bystander and whether under certain circumstances the civilian might yet be taken as a prisoner of war. Noninterference is, of course, a modern concept; for most of history, noncombatants were viewed as potential combatants and subject, therefore, to detention. For a discussion of the predatory aggression associated with premodern notions of war-making, see R. O'Connell, *Of Arms and Men: A History of War, Weapons and Aggression* (New York: Oxford University Press 1989). The issue of the rights of civilians and the moral guidelines governing the war convention as it applies to civilians is judiciously explored by Michael Walzer, *Just and Unjust Wars: A Moral Argument with Historical Illustrations* (New

York: Basic Books 1977); see, for example, the discussion on pages 144–7, and 160–75.

8 In 1919, when a resolution was brought before the House calling on the government to state more clearly its policy regarding the remaining interned enemy aliens, the Honourable Arthur Meighen, then minister of interior, noted that the designation "prisoner of war" applied equally to civilian enemy aliens and to merchant marines who had been transferred to Canada at the start of the war. In response to the question of whether they could be classified as prisoners of war, inferring *military* prisoners of war, he stated erroneously, "I do not think there is any different designation applicable." *Hansard*, 24 March 1919, 757.

9 For a discussion of the articles of the Geneva and Hague Conventions and the difficulties in carrying out the provisions of the protocols, see H.E. Belfield, "Treatment of Prisoners of War," *Transactions of the Grotius Society: Problems of Peace and War*, 9 (1923): 131–47.

10 For example, the American-Prussian Treaty of 1785 (the Treaty of Amity and Commerce), reaffirmed in 1828, pledged both contracting parties "to prevent the destruction of prisoners of war, by sending them into distant and inclement countries, or by crowding them into close and noxious places." T. Woolsley, "Treatment of Prisoners," *American Journal of International Law* 12, no. 1 (1918): 382.

11 On the British-German agreements of 1917–18, see G. Phillimore and H. Bellot, "Treatment of Prisoners of War," *Transactions of the Grotius Society: Problems of Peace and War* 5 (1919): 47–63.

12 Richard Speed, *Prisoners, Diplomats and the Great War*, 90; and Robert Jackson, *The Prisoners, 1914–1918* (London: Routledge 1989), 140.

13 G. Davis, "Prisoners of War in Twentieth-Century War Economies," *Journal of Contemporary History* 12 (1977): 628.

14 The situation in Germany was somewhat mixed. Although, as with the Allies, relatively smaller numbers were employed in 1914–15, by the end of 1916, 1.1 million Allied prisoners were at work in Germany. Some 630,000 were employed in agriculture; the remaining 340,000 were put to work on a variety of projects. Speed, *Prisoners, Diplomats and the Great War,* 76–7; D. Morton, *Silent Battle: Canadian Prisoners of War in Germany, 1914–1919* (Toronto: Lester 1992), 66–9; and R. Garret, *POW* (London: David and Charles Publishers 1981), 100–16.

15 Allied POW treatment in Germany was the subject of much debate and public speculation both during and immediately after the war. There is ample evidence to suggest that the life of prisoners of war was rigorous and hard and not without danger. Brutality was also an unfortunate consequence of imprisonment. But the greatest suffering endured it would seem was more a result of neglect due to the scale of the operations than a result of the routine use of coercion and punishment. Indeed, this also appears to be the case in Russia, where appalling rates of casualties among prisoners of war were recorded as a result of gross negligence. The issue of negligence, of course, raises the question of moral culpability, and in this sense the needless suffering and attendant losses speak loudly. Those who failed in their duty share the burden of responsibility. But that much having been said, a distinction focusing on intent, fine though it may be, can be made between a policy that openly suppressed prisoners' rights and a policy that ignored those rights. Speed, *Prisoners, Diplomats and the Great War,* 63–79 and 121–2; Garner, "Treatment of Enemy Aliens: Measures in Respect to Per-

sonal Liberty," *American Journal of International Law* 12, no. 1 (1918): 50–1. On the fate of prisoners of war in Russia, see G. Davis, "The Life of Prisoners of War in Russia, 1914–1921," in S. Williamson and P. Pastor, eds., *War and Society in East Central Europe,* vol. 5, *Essays on World War I* (New York: Brooklyn College Press 1983), 163–96. For an assessment of German use and treatment of Canadian POWs, see Morton, *Silent Battle*, 66–94; and J.F. Vance, *Objects of Concern: Canadian Prisoners of War through the Twentieth Century* (Vancouver: University of British Columbia Press 1994), 25–34.

16 For a discussion of this point, see Speed, *Prisoners, Diplomats and the Great War*, 183–9.

17 The issue of the length and character of the workday was raised on several occasions. J. Harkin, commissioner of the parks, communicated to the various park superintendents in the winter of 1916 that with the change in the season and the concomitant lengthening of the day, arrangements were to be made at each locale for prisoners to work longer hours. At Banff the time allotted for the march to the worksite became a subject of great complaint. Originally the daily march was not considered part of the workday, usually adding four to six hours to the daily work schedule. This was to change when, investigating conditions at the Banff camp, the commanding officer of Military District 13, General Cruikshank, observed that the thirteen-mile march to the worksite in the deep snow constituted a day's work in itself. NA, RG 84, vol. 124 file Y176, E.N. Russell, superintendent, Yoho and Glacier Parks, to J. Harkin, commissioner, Dominion Parks, 29 February 1916; and NA, RG 24, vol. 4721, file 2, Brigadier General E.A. Cruikshank to General Otter, 3 March 1916.

18 In reply to an enquiry from headquarters regarding the number of sick, the

commanding officer of the Edgewood Internment Camp reported that there were twenty internees who were pronounced incurable or tubercular. Their ailments ranged from curvature of the spine to rheumatism, hernias, cardiac debility, and senility. It was noted in each case that destitution was the cause of arrest. NA, RG 6 HI, vol. 752, file 3130, Captain C. Harvey, Edgewood Internment Camp to Major General Wm Otter, director, Internment Operations, 1 June 1916.

19 The official number of calories in a daily Canadian internment ration was calculated to be 2,595.52. The food allotment fell short of this total. Moreover, the energy required for a labourer was estimated at 2,903 calories. For a statement on the type and amount of rations provided, see B. Kordan and P. Melnycky, eds., *In the Shadow of the Rockies: Diary of the Castle Mountain Internment Camp, 1915–1917* (Edmonton: CIUS Press 1991), 18.

20 The undisciplined nature of the guards mirrored the rough treatment commonly meted out to the prisoners. Prisoners were beaten, pistol-whipped, shot at, and in some recorded instances pierced in the buttocks with bayonets. The attacks prompted reprimands from both General Otter and senior ranking officers, although warnings against brutality and unnecessary ill treatment appeared to have little effect. On the contrary, complaints lodged by the prisoners were often dismissed as petty and inconsequential. See NA, RG 24, vol. 4721, file 1, Colonel (later Brigadier General) E.A. Cruikshank, officer commanding, Military District No. 13 to Major General Wm Otter, director, Internment Operations, 16 November 1915; Major Duncan Stuart, commandant, Banff Internment Camp, to Col E.A. Cruikshank, 4 December 1915; and Brigadier General E.A. Cruikshank to Major Stuart, 20 December 1915. See also NA, RG 24, vol. 4721, file 3, "Summary of Prisoner

Complaints (Banff)," 10 December 1916; and NA, RG 6 HI, vol. 765, file 5294, C. Babyj, censor, Morrissey Internment Camp to Major General Wm Otter, director, Internment Operations, 1 September 1917.

21 When guards at the Banff Internment Camp were approached for service overseas, fully one-quarter of the entire company, who were described as "anxious volunteers," stepped forward. Because it was feared that the operation at Banff would suffer from the loss of so many men at once, only a few were discharged for active overseas duty. That decision prompted several guards, desperate to have done with the place, to plead for release from their current duty. Banff was not an exception, petitions by privates were regularly received by commanding officers of the various military districts. See NA, RG 24, vol. 4721, file 2, Col G. Moffit, 137th Overseas Battalion to Brigadier General E.A. Cruikshank, 17 February 1916; and Major Spence, commandant, Banff Internment Camp to General Cruikshank, 21 February 1916. For an example of a petition for release, see NA, RG 24, vol. 4721, file 2, Private G. Lomax to Brigadier General Cruikshank, 22 April 1916.

22 The insufficient rations were of some concern among those who were overseeing the work. It was observed by the Mara-Sicamous Road superintendent, R.W. Bruhn, that the rations were inadequate even for a man who was not working, let alone for a man who was expected to do hard labour. Voicing his frustration, Bruhn noted "The daily meat ration amounts to 5½¢ per man. I am certain this amount is lost several times over when the men are not working as they should." The lack of any response from Internment Operations to improve the situation prompted a threat from the deputy minister of public works "to close down the work – unless they are fed like human beings." See BC Ministry of Transportation (Freedom of Information

Branch), North Okanagan and Salmon Arm Districts (1915), file 211, R. Bruhn, assistant road superintendent, Mara-Sicamous Road, to J.E. Griffith, deputy minister, Department of Public Works, 21 January 1916; J.E. Griffith to Lieut. Colonel W. Ridgway-Wilson, Department of Alien Reservists (Esquimalt), 1 February 1916; and Col Ridgway-Wilson to J.E. Griffith, deputy minister, 4 February 1916.

23 On withholding payment and the administrative debate on the utility of this practice, see BC Ministry of Transportation (Freedom of Information Branch), North Okanagan and Salmon Arm Districts (1915), file 211, R.W. Bruhn, assistant road superintendent, Mara-Sicamous Road, to J.E. Griffith, deputy minister, Department of Public Works, 17 July 1916; R. Bruhn to J.E. Griffith, 10 August 1916; J.E. Griffith to Lieut Colonel W. Ridgway-Wilson, Department of Alien Reservists (Esquimalt), 17 August 1916; Major General Wm Otter, director, Internment Operations, to the Hon. T. Taylor, BC Minister of Public Works, 29 August 1916; and Bruhn to J.E. Griffith, 9 October 1916.

The preoccupation with expenditures would translate into several bizarre practices, not the least of which was the need to assess the cost of recapturing a prisoner. The expense of doing so was invariably to be weighed against the benefits to be derived from his labour. Other cost-saving measures included the use of two pairs of summer socks (they were available) as a substitute for a single pair of winter socks (they were yet to be purchased). See NA, RG 24, vol. 4729, file 3, Major Duncan Stuart, commandant, Banff Internment Camp, to Colonel E.A. Cruikshank, officer commanding, Military District No. 13, 28 October 1915; BC Ministry of Transportation (Freedom of Information Branch), Okanagan District (1915), file 1752, section 3, J. Black, road superintendent, Vernon-Edgewood Road, to J.E. Griffith, deputy minister, Department of Public Works, 22 November 1915; and BC Ministry of Transportation (Freedom of Information Branch), North Okanagan and Salmon Arm Districts (1915), file 211, R. Bruhn, assistant road superintendent, Mara-Sicamous Road, to J.E. Griffith, deputy minister, 25 December 1915.

24 NA, RG 13 B8, vol. 1368, file "July-October 1915," Secretary of State for Foreign Affairs to the United States Ambassador (Despatch No. 142861/15), 9 October 1915.

25 The difference would lessen as camps closed and Austro-Hungarian and German inmates were consolidated at certain locations. On the difference in treatment, see Morton, *The Canadian General*, 333; and Melnycky, "The Internment of Ukrainians in Canada," in F. Swyripa and J.H. Thompson, eds., *Loyalties in Conflict: Ukranians in Canada during the Great War* (Edmonton: CIUS Press 1983), 7–8.

26 A small, but poignant, example of this phenomenon was that according to the Hague regulations prisoners of war were to be paid the equivalent of the national army's standard rate of pay for the work performed. For work rendered outside normal military duties a soldier was paid a daily supplement of 25¢. That Austrian labourers were paid the supplementary rate, as opposed to the standard rate, underscored their nonmilitary status. Considered to be common labourers and noncombatants, they were provided with only the supplementary nonmilitary rate.

27 NA, RG 24, vol. 4721, file 3, Report of American Consul, "Visit to Internment Camp at Banff, Alberta," 25 May 1916; and United States National Archives and Record Administration, RG 59, file 763.72112/1958, Harold C. Clum, American Consul at Calgary, Alberta, to the Secretary of State, Washington, 25 May 1916.

The American Consul in Vancouver, J.C. Woodward, took a more sympathetic view of the plight of the internees, unofficially communicating to the inmates their rights under international law. It appears, however, that he made no formal protest against the conditions. See NA, RG 6, vol. 752, file 3033, J.C. Woodward, American Consul to unidentified prisoner, Mara Lake Internment Camp, 27 August 1916.

28 In September the weather turned wet and cold. There was snow on September 11th and rain frequently thereafter with snow also. On August 30th I asked for mackinaw coats, boots, better overalls, woollen mitts with leather pullovers, and pointed out that the weather got cold early here.

Throughout the latter part of September and October there were many complaints on account of the cold and the want of boots, mitts and overalls. On September 7th I asked to have clothing hurried forward. On September 30th I wired for clothing. During the latter two-thirds of September and in October the weather got colder. The prisoners were out of boots and overalls. September 11th I wired for blankets to be sent by express, but as stated above did not get them until October 2nd. On October 5th I asked particularly for mitts as the weather was turning cold. On October 13th I wired saying I badly needed overalls, socks, boots, mitts, and top shirts as ordered in September. On October 19th the outside mitts arrived, but no woollen ones. On October 21st I wired for permission to purchase some boots and overalls locally. On October 27th and November lst I again wired for this permission saying I could not order men to work in snow and slush without reasonable clothing.

During October the prisoners complained bitterly of want of boots and overalls, and of having to sleep in tents without fires. One prisoner who escaped left a letter to a friend saying he preferred to take the chance of being shot escaping than to live under the conditions. (NA, RG 24, vol. 4729, file 3, Major Duncan Stuart, commandant, Castle Internment Camp, to Colonel E.A. Cruikshank, commanding officer, Military District No. 13, 4 November 1915)

29 E. Satow, "The Treatment of Enemy Aliens," *Transactions of the Grotius Society: Problems of Peace and War* 2 (1916): 8.

30 P. Panayi, "Germans in Britain During the First World War," *Historical Research* 44 (February 1991): 63–76.

31 Speed, *Prisoners, Diplomats and the Great War*, 141–53.

32 For a remarkably detailed and neutral account of civilian enemy alien life under German authority, see J.D. Ketchum, *Ruhleben: A Prison Camp Society* (Toronto: University of Toronto Press 1965).

33 J. Garner, "Treatment of Enemy Aliens: Right of Access to the Courts," *American Journal of International Law* 13, no. 1 (1919): 22–59.

CHAPTER FIVE

1 *Annual Report of the Department of the Interior, 1916* (Ottawa: King's Printer 1917), 6–9.

2 Desmond Morton, *The Canadian General: Sir William Otter* (Toronto: Hakkert 1974), 331. At the going daily rate of 25¢ Internment Operations accounts would show that at Kapuskasing, for 6,145 man-days of work provided by alien labour during the month of February 1916, the Department of Lands paid out $1,536.25. National Archives of Canada (NA), RG 6, vol. 760, file 4978, "The Ontario Government in Account with Internment Operations."

3 Unlike other provincial governments, the British Columbia government took an active interest in the internment of enemy aliens, acquiring, for instance, the services of Major W. Ridgway-Wilson to act as the province's chief internment officer, heading up a so-called Department of Alien Reservists. Wilson was taken on by Major General Wm Otter as a staff officer with the Internment Directorate, assuming direct responsibility for the camps in Military District No. 11 (British Columbia). NA, RG 24, vol. 851, file HQ 54-21-11-14.
4 As in the case of the Department of Lands in Ontario, the BC Public Works Department agreed to remit payment for the alien prisoners to Internment Operations. NA, RG 6, vol. 760, file 4978, Major General Otter to the minister of lands, government of Ontario, 18 February 1916.
5 The cited numbers of prisoners are calculated from reports contained in NA, RG 117, vol. 20, file "Sundries re: All Camp," "Movements of Prisoners of War."
6 See, among others, BC Ministry of Transportation (Freedom of Information Branch), Okanagan District (1915), file 1752, section 2, J.W. Jones, Mayor, City of Kelowna to the Hon. T. Taylor, BC Minister of Public Works, 4 September 1915; and Taylor to Jones, 24 September 1915.
7 BC Ministry of Transportation (Freedom of Information Branch), Okanagan District (1916), file 1752, section 4, J. Black, superintendent, Vernon-Edgewood Road to J.E. Griffith, deputy minister, BC Department of Public Works, 31 August 1916; and the Hon. T. Taylor, BC Minister of Public Works, to J.E. Griffith, deputy minister, Department of Public Works, 2 September 1916.
8 I beg to acknowledge receipt of your favour of the 4th instant, submitting

a resolution unanimously passed by the citizens of Enderby at a representative meeting held on the 27th ultimo, and in regard to the employment of interned aliens on the Revelstoke to Vernon Road between Sicamous and Mara.

I may advise you in reply that, with the consent of Major General Sir Wm Otter, the Officer Commanding Internment Operations in Ottawa, the Public Works Department has been allowed to use some two hundred interned aliens on the Vernon-Edgewood Road. Since this permission was granted, a great many applications have been forwarded to this Department suggesting the use of alien prisoners on other roads in the province. All such applications have been placed before the authorities in Ottawa. The applications are so many throughout the province that, apparently, Major General Otter is not in a position to act on them immediately. Furthermore, I am just in receipt of a telegram from him to the effect that Colonel Macpherson, his Staff Officer, is on his way to Victoria, and that he will visit me in the course of a few days for the purpose of looking into these many applications. I shall be very glad to advise you after consulting with him as to whether the wishes of your citizens can be met. (BC Ministry of Transportation (Freedom of Information Branch), North Okanagan and Salmon Arm District (1915), file 211, Hon. T. Taylor, BC Minister Public Works, to Enderby and District Conservative Association, 8 June 1915)

9 At a meeting of the District Conservative Association held at the Fernie hotel on Wednesday afternoon, the following resolution was passed: WHEREAS there are at present in Fernie from 350 to 400 alien enemies to be interned within the next 24 hours and this number will undoubtedly

be added to in the near future; AND WHEREAS the cost of transportation for every man from here to the coast would be in the neighbourhood of $30 per man; AND WHEREAS there is a deserted town-site owned by the Crow's Nest Pass Coal Company Limited with buildings, waterworks, and necessary outbuildings capable of accommodating over a thousand persons which could be acquired immediately; THEREFORE BE IT RESOLVED that this meeting request the proper authorities to make the necessary arrangements to secure the Morrissey Mines town-site as an internment camp for this district. (*Fernie Free Press*, 11 June 1915)

10 Morton, *The Canadian General*, 341.

11 NA, RG 84, vol. 124, file Y170, J.B. Harkin, Commissioner of Parks, to General Otter, 28 September 1915. Harkin would describe the arrangement as follows in his annual report: "It was felt that it was not good for the prisoners to live for months in a state of idleness; that it would be advantageous for them to have work to do and that having to maintain them in any case it would be good business for the Government to secure with such labour the construction of roads and other public works in the park." *Annual Report of the Department of the Interior, 1916*, 4.

12 Canada, House of Commons, *Hansard*, 15 February 1916, 849.

13 See Morton, *The Canadian General*, 338. The Amherst station and the citadel in Halifax were reserved from the start for bona fide prisoners of war – foreign nationals, namely merchant seamen, captured elsewhere and interned in Canada at the request of Imperial authorities. Fort Henry and the Vernon camps were emptied of their non-German inmates in 1916. The camp created at Morrissey was initially to hold only "Austrians," but 143 enemy aliens of German nationality were sent from Brandon and Lethbridge after these two camps had been dismantled and the remaining pris-

oners dispersed. Yet even within this camp setting, class and ethnic differences were observed: the large hotel structure was reserved for the East Europeans, while a second, smaller building, especially built, was set aside for first-class prisoners. See plan of compound, NA, RG 24, vol. 4661, file 99-237 (1), "Court of Enquiry," 27 November 1916.

14 The internment experience at Castle Mountain is documented in B. Kordan and P. Melnycky, eds., *In the Shadow of the Rockies: Diary of the Castle Mountain Internment Camp, 1915–1917* (Edmonton: CIUS Press 1991). An account of the Castle Mountain Camp and detailed sketches of internment camps in Jasper, Revelstoke, and Yoho national parks are also to be found in B. Waiser, *Park Prisoners: The Untold Story of Canada's National Parks, 1915–1946* (Saskatoon and Calgary: Fifth House Publishers 1995), 3–47.

15 NA, RG 84, vol. 124, file Y170, J.B. Harkin, Commissioner of Parks, to General Otter, 28 September 1915.

16 "Commissioner of Parks to Arrange for Camp," *Revelstoke Mail Herald*, 28 July 1915.

17 WHEREAS the Dominion Government saw fit to establish a park on Mount Revelstoke and created an internment camp last year for the purpose of completing an auto road on the same but later moved the prisoners before the road was completed; AND WHEREAS it has been learned that for some unknown reason the government does not intend again to establish a camp immediately for the purpose of completing the road, which will be of no assistance to the city unless completed, and which, if the work is to be efficiently and economically done, must be finished during the months of May, June, July, August and September; AND WHEREAS the City of Revelstoke and District has contributed to His Majesty's service approximately eight hundred men and the war conditions have made the

up-keep of the city a heavy burden for the remaining number; BE IT THEREFORE RESOLVED that the City Council and other local bodies and organizations representing the entire citizens of Revelstoke protest against the lack of assistance on the part of the Dominion Government and military authorities in not returning here the prisoners and completing the said road at once; That the City of Revelstoke ask to be treated on the same basis and with the same concessions as some other places no more deserving; That the Dominion Government and the military authorities be asked to re-establish an internment camp at once so that the work may be completed during the present summer; AND BE IT FURTHER RESOLVED that if there is any reason why the establishing of a camp and the completion of the road on Mount Revelstoke is not feasible through expense or otherwise, the citizens of Revelstoke be made acquainted with such reason, so that the matter can be investigated and the return of the camp be at once affected. (NA, RG 84, vol. 190, file MR 176, Municipal Resolution, re: Re-establishment of Revelstoke Internment Camp)

18 See NA, RG 84, vol. 124, file Y176, E.N. Russell, superintendent, Yoho and Glacier Parks, to J.B. Harkin, commissioner, Dominion Parks, 9 August 1916; Lieutenant G.H. Brock, commanding officer, Otter Internment Camp, to E.N. Russell, superintendent, Yoho and Galcier Parks, 12 August 1916; E.N. Russell to J.B. Harkin, 12 August 1916; J.B. Harkin to E.N. Russell, 22 August 1916; and Lieutenant G.H. Brock, commanding officer, Otter Internment Camp to Major General Wm Otter, Director, Internment Operations, 23 August 1916.

19 Sir William Otter, *Internment Operations, 1914–20* (Ottawa: King's Printer 1921), 16.

20 The experience was a disappointment to the degree that the work per-

formed by the internees did not fulfill the expectations of senior park officials. But this sentiment was born out of frustration that in fact more could have been accomplished had "more stringent measures been resorted to." Bill Waiser, for example, writes, "As for the Dominion Parks Service, it looked back upon the internment camp experience as something of a mixed blessing. Parks officials had expected great things from the internees and were disappointed by what had been achieved in the end. They believed that the prisoners should have been forced to toil away – regardless of the circumstances and conditions – and resented not only the slow pace but also the prolonged work stoppages." Waiser, *Park Prisoners*, 47.

21 *Annual Report of the Department of the Interior, 1918* (Ottawa: King's Printer 1919), 6.

22 Ibid.

23 Whyte Museum of the Canadian Rockies, Accn. no. 1838, Interview, Colonel J. Anderson-Wilson, 4 May 1973.

CHAPTER SIX

1 Responding to the applications, officials prepared lists of candidates eligible for release. Upon inspection company representatives would select those who met their employment needs from the lists. Commanding officers, however, were granted the discretionary power to retain any individual. See National Archives of Canada (NA), RG 6 HI, vol. 755, file 3326 (iii), Directorate of Internment Operations to Major P.M. Spence, commandant, Banff Internment Camp, 7 April 1917.

2 Records pertaining to internment at Munson and Eaton are located in NA, RG 6 HI, vol. 768, file H6342.

3 The debate was touched off by several motions relating to an order paper dealing with alien labour in Canada that was put before the House. It was argued that there was a need for all regulations and orders in council governing alien labour in Canada to be consolidated in one uniform law but that "in framing such legislation due regard should be had to the local conditions and needs of the different parts of the Dominion." In effect, the motion before Parliament called upon the legislative body to acknowledge and reaffirm the policy of internment. The discussion, meanwhile, centred on the necessity and utility of internment and its possible expanded use. References are drawn directly from Canada, House of Commons, *Hansard*, 22 April 1918, 973–1025.

4 In point of fact, Japanese nationals who had immigrated to Canada before the war were serving in the Canadian Expeditionary Force. See Roy Ito, *We Went to War: The Story of the Japanese Canadians Who Served during the First and Second World Wars* (Stittsville, ON: Canada's Wings 1984).

5 From 1915, the standard government of Canada reply to German diplomatic protests regarding the use of German prisoner labour was "that all work performed by prisoners of war at Canadian internment stations is quite voluntary and no pressure is brought to bear upon them." By late 1917 the spate of consular reports providing evidence to the contrary highlighted the increasingly untenable nature of the official position of the Canadian government. See NA, RG 6 HI, vol. 819, file 2616 – "Downing Street: Compulsory Labour and Prisoners of War."

APPENDIX

Prisoners' Rolls

Castle Mountain/Banff
Revelstoke/Otter
Monashee/Edgewood
Jasper Park
Mara Lake
Morrissey

CASTLE MOUNTAIN/BANFF

NAME	NUMBER
Adam, M.	61
Adamus, M. (Wawzyniec)	219
Andrejciw, John	361
Andrusak, Harry	611
Antoniuk, Fedor	416
Antoshiw, John	538
Augustinowycz, Paul	1
Babiuk, Zachary	373
Bablick, J.	128
Babyn, Alex	
Bachinski, Joe	410
Baczynski, Michael	2
Baida, Harry	62
Balacz, Gabor	546
Balas, Hrynkow	380
Balun, John	295
Baran, Luka	573
Baran, Tom	570
Bardko, Petro	579
Barkow, Wasyl	272
Baron, S.	294
Barth, Adam	264
Beloski, Alex	426

Prisoners leaving for Castle Mountain, Banff

Belouce, Alex	548
Bentryn, Mike	516
Berezka, M.	127
Berezuk, John	259
Berlacz, Marko	126
Berold, Pete	522
Bertolini, Candido	484
Bezkorowai, K.	191
Bidniak, Jakob	125
Bignotti, Vitorrio	483
Bilinski, Josef	244
Biyk, Alex	329
Blyzniuk, Nick	65
Bodnaryk, Wasyl	287
Bogdanovitch, Michael	288
Bogu, Aron Tymac	123

Boresky, John	402
Borowczuk, F.	396
Borszcz, Wasyl	633
Bortnik, Bill	320
Bosalenka, John	543
Bota, Constantin	206
Brad, Stefan	618
Bronikowski, Frank	493
Brown, J.	122
Buchalski, John	240
Bucij, Fred	513
Buckanoko, Antonio	443
Budak, George Luka	631
Bugera, Roman	326
Buis, Alex	525
Bullak, Pete	431
Buzenski, P.	121
Byn, Mike	399
Bysko, M.	648
Camelback, John	580
Camysar, Nykola	529
Cermalish, Petro	129
Chebrak, Nikola	360
Chechul, N.	131
Chemereky, Wasyl	328
Cherbanuk, John	365
Cherys, Andrey	563
Chiko, Harry	268
Chorkulak, Andrew	130
Chornoby, John	504
Chorny, Oleksa	379
Chustinsky, S.	303
Cierea, Louis	625
Cook, Louis	531
Cooper, John	207
Cyhanczuk, Denis	196
Czepil, Mike	434
Czervinsky, Tony	369
Czornyj, Fred	481
Dabos, Stefan	606
Danyliuk, Alek	302
Daniliuk, Pete	351
Danyluk, D.	132
Danyluk, M.	7
Danyluk, Mike	419
Danylyski, Wasyl	249
Demczuk, N. (Czeladyn)	133
Deminczuk, John	535
Denchek, Sam	347
Deputat, Nick	197

Gojan, Bill	71
Gontowluk, Luke	70
Goriuk, Nick	10
Gowrylik, John	348
Gozman, Stefan	307
Grasmajar, Ignat	414
Greenvold, Z.	638
Gregorijczuk, J.	415
Gregovich, Alexa	319
Gubransky, John	595
Guzumaniuk, J.	138
Haftkowicz, A.	253
Haida, Metro	442
Hamschuk, Prokop	308
Handyburak, G.	140
Hansivicz, Steve	377
Harasymiuk, P.	74
Hawryliuk, J.	73
Hawryliuk, Mike	475

Returning to camp, Castle Mountain

Hawrysynyzn, J.	12
Hepp, A.	651
Heuchert, Johan	565
Hirocks, Pete	336
Hnatkow, Sam	569
Holaryk, George	296
Holowatyj, Petro	508
Horodaski, John	266
Horodenski, Pavlo	327
Howrylkow, Mike	556
Hryhercyk, Alec	514
Hryhorchuk, Wasyl	549
Huculiak, Wasyl	228
Hudash, Tony	418
Hudiak, Alek	446
Hudyma, Mike	422
Hulinga, Nestor	617
Hulonga, Agapi	75
Humeniuk, John	324
Huminiuk, John	13
Ic, Joseph	445
Ihnatiuk, Nykola	540
Ivesach, G.	78
Ivesach, P.	77
Iwanychuk, Mike	8
Iwasieczko, Michael	76
Jablonski, Josef	515
Jarabos, Mike	199
Jarmoli, B.	427
Jastrybczak, Mike	392
Jaworenko, Petro	235
Jelanik, Joe (Lalik, M.)	152
Jelanik, John	79
Jelonick, J.	14
Jeremczuk, John	258
Jubas, Louis	619
Jucz, Wasyl	80
Jurkowski, Nick	463
Kagenovitch, D.	236
Kalyniuk, Dmytro	567
Kamenuk, N. (Olynyk)	141
Kaminski, Nick	355
Karpiecz, Anton	
Karpiuk, S.	273
Karsczisze, J.	498
Karuk, Nykolaj	84
Kash, Mike	271
Kash, Mike	314
Kasmer, Martin	229
Kasovitch, Wasyl	626
Katchuk, Steve	278
Kato, Billy	544

Name	No.
Kempe, John Carl	597
Kerczuk, John	420
Kibik, Bill	466
Kinish, Bill	86
Kis, Johann	596
Kitzmantel, Mike	193
Klimczuk, G.	148
Klym, Kost	338
Klym, Mike	600
Koboski, Stanislaw	587
Kobylczuk, Josef	612
Koczur, S. (Tkaczyk, D.)	20
Koenig, Joseph	265
Kola, P.	349
Kolasky, Dmytro	334
Kolej, Petro	226
Kolenzuk, Mike	576
Koliaska, Nick	82
Koma, Mike	629
Komarnitski, Kasper	21
Komhyr, Dmytro	97
Kondro, John	224
Konoski, Kost	341
Konowalczuk, Pete	290
Konstantiniuk, Dmytro	221
Korkowsky, Mike	267
Kosewan, Alex	604
Koshkur, Ilko	437
Kostyniuk, G.	19
Koszkor, Paul	519
Koszyl, Josef	507
Kotowicz, Pete	632
Kowalyk, Steve	591
Kowayk, Nick	585
Koza, Joseph	557
Kozach, J.	243
Kozak, Mike	511
Kozar, V.	263
Koziaz, Peter	18
Kozma, Nykolai	233
Koznia, Kirylo	439
Krajasow, George	15
Krasneski, Joe	588
Krasy, Mike	374
Krawec, J.	209
Krewecki, M.	398
Krilovitch, P.	150
Krisman, M.	88
Krupka, Nick	641
Krymniuk, Andro	149
Kubel, Joe	578
Kucuper, Jan	417
Kuczyk, O.	146
Kukoy, Charles	636

Name	No.
Kuley, Stefani	47
Kulowechuk, Mike	274
Kulyk, J.	438
Kumpf, Frank	560
Kupchenko, John	523
Kuriyk, Harry	276
Kuryliuk, Dmytro	83
Kushniruk, John	453
Kuszner, I.	142
Kutybaba, Fred	425
Kuz, Steve	401
Kuz, Wasyl	400
Kuzio, G.	252
Kuzio, N.	17
Kuzio, N.	81
Kuznier, Simeon	452
Kwasny, Albert	87
Kwasny, Jan	22
Kwasny, J.	144
Kwasny, M.	145
Kwasny, Tom	630
Laba, Andrew	151
Laban, S.	359
Laneski, Joseph	275
Lanowski, Mike	23
Lapsuliak, Wasyl	382
Lepecky, Dmytro	551
Leskiw, Joseph	
Lewicki, Danylo	90
Lipka, H.	344
Lizanski, Wojtko	506
Lobay, John	610
Lucak, D.	25
Luciw, Basil	552
Luczka, Andrey	472
Lukasiewich, Pete	581
Lukitski, Frank	478
Lukitski, John	477
Lytwyn, John	602
Madrigor, Dmytro	447
Makar, P.	29
Makichuk, Mike	499
Maksymic, Mike	502
Malena, Augustin	643
Malkus, J.	156
Mandich, I. (Karanovich)	622
Mandiuk, Andrey	332
Mandrik, Michael	159
March, Bill	280
Marchuk, John	590

Marchuk, Philip	589
Marek, M.	26
Marentiuk, J.	386
Mark, Fred	564
Marowski, Pete	539
Marteniuk, John	366
Marunczak, John	200
Masniuk, Peter	28
Matheas, F.	161
Melnichuk, F.	364
Melnik, Andre	293
Melnyczuk, D.	164
Melnyk, Nick	652
Merowski, Michael	512
Mescowski, Geo.	91
Mezner, J.	160
Mibroda, P.	93
Micenko, Harry	158
Mickowic, Tarko	155
Mike, Nick	583
Mikline, Andrew	451
Milak, Steve	95
Milkovitch, Nikola	509
Miller, Edward	217
Miller, Jan	609
Miller, Mike	208
Milon, Mike	408
Mintanko, Steve	340
Mirwald, Joseph	520
Mishchop, Andrew	353
Misinczuk, P.	412
Mizyra, F.	494
Modunal, J.	230
Moroz, Alex	343
Morris, Frank	460
Morris, Louis	92
Morris, Steve	162
Mostowyj, Mike	195
Mudry, N.	157
Mueritz, Bill	574
Myroniuk, H.	432
Mysk, Nick	299
Mysz, Lazo	163
Mytnyk, Wasyl	317
Napadi, P.	165
Nazar, Mike (Botojuk, P.)	635
Negrych, Nick	640
Neisher, Harry	356
Neowsan, Ivan	616
Netrebiak, D.	378
Niemsov, Joe	532

Polij, Paul	489
Poliski, Joseph	248
Pololaychuk, Harry	306
Postinuk, John	300
Posyniak, M.	169
Povlichuk, Kendrat	311
Preglar, Stefan	545
Przybyla, J.	220
Przybylak, Joseph	440
Puhacz, A.	388
Pujniak, Wasyl	405
Pulady, Frank	238
Purdich, George	620
Pusyniak, George	448
Radmaka, Fredrick	603
Ragoszewski, Louis	413
Rakoski, Nikola	592
Reder, Frank	568
Relian, Pete	421
Repeteski, Mike	541
Retecky, Mike	429
Rewega, Joe	465
Rilszeski, Nykola	381
Roberts, David	105
Roman, Mike	255
Romaniuk, George	501
Romaniuk, J.	575
Romaniuk, Roman	488
Romanovich, Ostap	387
Ronyke, Alex	582
Rosbosky, Onufry	524
Roska, Tom	38
Rosky, Mike	455
Roszakowski, N.	201
Roth, Mike	395
Rovnczka, Philip	479
Rubiaski, Klemens	171
Rudik, P.	104
Rupustka, T.	301
Ruski, Mike	461
Rybuch, J.	204
Rypka, Andrew	103
Sadak, Valant	48
Salie, Wasyl	526
Sankiw, B. (Grech, E.)	11
Sapiak, John	45
Sawchuk, Bill	457
Sawchuk, John	279
Schaaf,	645
Scrypnik, John	397

Name	No.
Seidl, Anton	644
Semkiw, Mike	385
Sereka, Tony	352
Sesk, Joseph	24
Shapka, P.	44
Shernuzan, George	270
Shumowsky, E.	384
Shust, Harry	593
Shutts, M. (Chalaturnyk)	175
Shuty, George	550
Shymanski, A.	521
Simichuk, Fred	316
Sinuk, Harry	335
Skaradiuk, J.	40
Skomorowsky, Bill	330
Skoropodak, Mike	281
Skrynyk, Stanley	337
Skyaban, John	110

Prisoners' compound, northeast view, Castle Mountain

Skypnyzuk, D.	41
Slesvige, Voitko	283
Smith, John	482
Smith, Mike	471
Smoliy, D.	315
Sniatynski, John	172
Sobczuk, Pawel	459
Sobota, Steve	393
Sokirka, Harry	358
Sokolowski, Mike	350
Solinski, Artymon	608
Sopkiw, R.	177
Sorochty, Mike	594
Soville, L.	179
Sowyk, F.	198
Stachera, Dmytro	607
Stackwell, Mike	282
Stakiw, U.	284
Stankowski, Nick	192
Starzeinski, Joe	428
Stefanchuk, Wasyl	370
Stefanesol, Joseph	615
Stefaniuk, F.	39
Stefanko, J.	473
Stefanovich, Dmytro	310
Stefanyszyn, N.	176
Stefiuk, Wasyl	222
Sterko, Mike	325
Sternal, Albert	173
Stillin, John	500
Subat, Andrew	46
Sugak, Mike	433
Suhan, W.	251
Sukowich, Mike	178
Susak, John	577
Suson, Louis	49
Swerdlyk, Nik	339
Swityk, Harry	469
Swylak, P. (Serybajke)	50
Symotiuk, Andrew	109
Synyszyn, Dmytro	47
Szabatura, Ivan	342
Szcerebak, Steve	112
Szemlej, Mike	467
Szeremeta, K.	503
Szewczuk, Fred	239
Szopotiuk, Roman	436
Szostak, Hnat	174
Szuster, F.	490
Szydlowski, A.	42
Szydlowski, J.	107
Szymanski, Nick	371
Szymiec, S.	108

Tanasiuk, J.	357
Tani, Alec	449
Tanovich, Alex	456
Tatchuk, T.	318
Tawaski, Tankow	554
Teley, Andy	55
Tepic, John	182
Tesliuk, Wasyl	257
Thachuk, Nykola	424
Tiureau, John	623
Tkachuk, Peter	183
Tkachyk, Wasyl	277
Tkaczuk, J.	250
Tkaczuk, Nick	642
Toderiuk, Alex	517
Todoziczuk, Dan	285
Tomak, N.	404
Tomasz, M.	184
Tomkovich, E.	215
Tomkovich, M.	214
Topolnicki, J.	614
Torebalk, George	322
Tosta, Steve	527
Toth, John	51
Tressenberg, Joseph	584
Trulik, Dmytro	52
Tryhuk, Andrey	180
Trypczuk, William	487
Tushkow, Adam	368
Tuski, Mike	269
Twanyk, Luka	261
Twaranski, M.	114
Tyczynski, J.	113
Tymczuk, John	476
Typluk, P.	181
Tysliuk, Wasyl	468
Ukrainyc, Mike	605
Umritz, Bill	574
Urbaniec, Anton	53
Urdea, John	624
Vad, Charles	571
Vamunov, Saro	627
Vechman, Pete	533
Vishovski, Steve	312
Voida, Mike	313
Vyslisky, Joseph	407
Wala, S.	188
Wartyk, Mike	558
Wasiluk, Dan	430

Wasywich, Nykolaj	245
Weiss, W. A.	639
Weleszczuk, George	117
Werski, Mike	450
Wesylenchik, John	454
Wewcheruk, Illa	116
Wewcheruk, Wasyl	115
Winski, Tom	189
Wintz, Louis	185
Wodianchuk, Dmytro	485
Woityshen, M.	54
Wolchuk, John	435
Woytowich, Emil	362
Wynnyczuk, Nykolai	234
Wynnyczuk, Steve	203
Yustyak, Mike	118
Yvanoff, Bill	213
Zablotski, N.	120
Zagursky, H.	345
Zahara, Mike	653
Zalenchuk, Mitro	389
Zalewski, F.	354
Zapoticznyj, J. (Zik)	59
Zarycki, John	56
Zarytoshchia, J.	375
Zazula, F.	60
Zelenko, W.	231
Zewicky, Dan	89
Ziawin, A.	119
Zice, Harry	376
Zinkiewicz, D.	187
Zowtiak, John	58
Zowtiak, Michael	186
Zuleski, S.	599
Zulyk, M.	153
Zychowski, Alex	57

REVELSTOKE/OTTER

NAME	NUMBER
Adam, Wasyl	
Andruch, George	
Bachinski, Mike	2
Baychick, Andrew	81
Bersan, Nikolai	
Boy, George	130
Cencic, Joe	92
Cerha, Antoni	136
Crucic, Joseph	

Sentry watch, Otter

Lafsansky, George	55
Marchuk, Harry	
Morosan, Teodor	
Muntgen, Peter	184
Ostafiw, Roman	
Pavelich, Ivan	28
Petnik, George	
Polici, E.	66
Radich, Joe	70
Rozmo, Kazik	77
Sakish, Mike	19
Shingara, Fred	51
Suranczuk, Peter	
Urdea, John	
Vanmov, S.	
Velyko, Ivan	
Vidas, John	93
Vitalich, Lazar	
Volny, John	89
Yukas, Louis	
Zinzar, Stanislaw	

MONASHEE/EDGEWOOD

NAME	NUMBER
Alivoprodic, Anton	682
Barbuletin, Vasyly	
Belanko, S.	5
Bonawink, Mike	414
Bonora, Joe	462
Calusic, F.	679
Cannock, Louis	807
Carnich, Louis	461
Carpi, Nick	442
Chichi, Flesk	438
Chrest, Vasyl	566
Coslyz, Rudolph	562
Crusch, Anton	805
Domaneg, G.	657
Drohobycki, John	439
Evasiuk, Steve	1

Fordigian, Baronur 78
Frager, Mike 139
Fritz, G. 356
Furchak, P. 603

Gabrille, George 524
Galazin, L. 512
Groch, Tom 694

Hanschuk, John 501
Harri, Jon 599
Hlachysh, L. 413
Hryncyshyn, W. 251
Hudem, M. 432
Hudina, W. 769

Ivanczuk, Ivan 700

Jawarski, J. 535

Kalapza, A. 420
Kapaliuch, M. 675
Kobray, M. 416
Kornopik, George 418
Korzie, Paul 422
Kulizycke, J. 406
Kupunjar, S. 579

Lachen, M. 533
Latven, Peter 774
Lazenko, Tony 606
Lukanchi, Tony 593
Lutyk, George

Manko, Fred 476
Max, H. 475
Maybroda, John 482
Mutich, Mike 536
Mykhaylyshyn, M. 381

Najas, Andrew 477

Panewiuk, Mike
Panjazyski, J. 48
Petraschuk, Steve 417
Petrovich, Steve 597
Phillips, Joe 592
Potpala, Mike 818
Potubaluo, A. 410
Protopaulim, M. 430

Radovich, Nick 577
Ruaban, Enrico 464
Rukavina, Peter 691
Rysink, H. 564

Sikora, Joe 558
Skolovz, Mike 527
Sofronu, Vladimir 429
Stanuschak, J. 501
Stefanos, Joe 690
Stepanik, Frank 465
Stepko, F. 583
Stoak, Mike 529
Stocki, Joe 531
Stowkowy, Mike 695
Studyma, M. 426
Stypeniuck, Lazar 6

Tomase, Angelo 453
Tomka, Tony 608
Totram, T. 498
Tuka, John 469
Twanczuk, J. 700

Cutting down the old road, Edgewood

Uzelac, Nick	539
Valorez, Leon	559
Wagner, Richard	510
Wasgisch, P.	159
Wisinsky, Fred	3
Zanoni, Pietro	450
Zurawick, Joseph	
Zurovich, Henry	
Zusitch, Ily	216

JASPER PARK

NAME	NUMBER
Andrako, Joe	92
Aparek, Ola	105
Babiocz, Francis	181
Bacal, Antoni	34
Baczynski, John	156
Balawayka, John	66
Baratensky, Joseph	83
Barysko, Steve	2
Belasky, Mike	89
Berchuk, John	186
Bialkoosky, Joe	121
Bidnsky, Vladmir	176
Bogrovich, John	29
Boraiko, Wasyl	141
Borcz, Jacob	12
Bross, Peter	15
Chernish, Mike	114
Chernoski, Metro	
Chewn, Stefan	177
Chislicki, Mike	
Chometa, Simko	170
Chowata, Fred	200
Chujka, George	116
Chuwinski,Tony	31
Ciubra, Jacob	
Czerniucizan, A.	203
Daczuk, Mytro	
Demkovich, Steven	206
Derkich, Alek	154
Dobta, Jacob	112
Douil, Mike	90
Dubener, Anton	151
Farce, Boisik	173
Feschuk, Yuri	

Enemy aliens at work on town project, Jasper Park

Galat, Peter	61
Galat, Roman	102
Gordika, George	88
Gorinczuk, Lazar	158
Gus, Stefan (Dzus)	160
Gulka, Onufry	131
Hadzik, Andy	10
Jarinchuk, Mich	190
Kagorsky, Stan	14
Katchur, John	125
Kizimir, George	184
Kolbuck, John	
Kopok, Nick	64
Koronackj, Wasyl	67
Korshiwski, John	145
Kosmoluk, Fred	143
Kovalchuk, Harry	155
Kozaczuk, John	165

Kulaj, Michael	37
Kusck, Roman	11
Maki, Lorenz	24
Makulak, Wasyl	85
Mandry, Jarema	1
Manihszyn, Mike	157
Marynski, Nic	
Mazar, Bazyl	119
Menitchuk, Jack	113
Milan, Ivan	30
Mochanowsky, Frank	130
Mochanowsky, Theo	129
Mortuk, Peter	51
Nawrocki, Tom	132
Novchuk, Peter	56
Ogilski, Anton	191
Pakulak, John	75
Pankovich, Frank	19
Paskowski, Antoni	101
Petrik, Nikola	55
Poroschuk, Turko	61
Prelipcean, Gabriel	84
Pylowsky, Nick	168
Pyziak, Tony	13
Rachy, Tom	179
Reshak, Joseph	182
Riska, C.	120
Rispeler, August	45
Rispeler, Charles	16
Roshin, Alexa	82
Rudie, Peter	167
Sanchuk, Onufry	95
Schmidt, Philip	149
Selloum, Mike	97
Seninsfika, Mike	21
Shewaga, Dmytro	70
Shewaga, Mike	71
Shumerbaroda, George	166
Shumovsky, E.	
Smylakovski, Petro	79
Stimac, Peter	208
Swall, Harry	153
Turck, Martin	39
Visucuk, Doan	124

Walla, Joe	46
Wasilewski, Walter	42
Wasiliuk, Wasyl	73
Wasiluck, Dmytro	196
Weevick, Jack	100
Wilikensky, Harry	152
Wylyaka, John	189
Yankevitch, Steve	44
Yorkowski, S.	
Yuriczuk, John	106
Zach, Ignatz	33
Zagan, Nikola	72
Zanak, Frank	54
Zdobolak, Joseph	
Zylawa, Harry	180

MARA LAKE

NAME	NUMBER
Adzijer, Sam	508
Andrijasevich, Mike	494
Antoniuk, Joe	270
Anurnik, Alex	224
Asluck, George	165
Babich, Mike	131
Balenovich, Louis	101
Basarab, Joe	204
Benik, John	145
Begar, Joe	184
Ber, Mike	304
Berg, Steve	234
Bezpilko, Nick	
Bilos, Anton	133
Biskufovich, Turay	99
Bizy, Wm	221
Blamish, Nick	169
Bobby, Tom	33
Bodmanovic, Joe	322
Bodnar, John	276
Bolaski, Mike	212
Boroja, Philip	178
Borro, Charles	186
Bovich, Nick	259
Broche, Rudolph	545
Broder, George	287
Brodnovitch, Mike	264
Burich, Joe	198
Caranovich, Mike	168
Cashuk, O.	

Name	Number
Charuiths, John	158
Chopiuk, D.	240
Chychul, Mike	253
Cirkovich, Issac	1
Citulski, Nicholas	620
Crolich, Mark	176
Cussick, Mike	116
Dacz, John	650
Danmill, Steve	146
Derick, Jacob	205
Dick, Sam	686
Diduch, J.	160
Dimitrasinobic, Mike	311
Dirana, John	102
Doskoch, Wasyl	385
Drebit, John	319
Dreksler, Joe	325
Drenavic, Pete	115
Drobach, Steve	195

Returning to prison compound, Mara Lake

Dudar, Alec	282
Duncan, Alex	206
Dvorsky, G.	792
Elko, Sam	43
Engil, Tom	201
Fedracky, Vasyl	278
Foderich, John	268
Forster, Mike	247
Fredilikh, Onufry	194
Galevich, Pete	226
Gerkenez, Peter	143
Grabar, Steve	179
Gravsik, Sam	543
Gubek, John	274
Harpuk, George	137
Haruck, George	161
Hidukovich, Javo	109
Hnidka, Joe	330
Hornoff, Emil	87
Hryniw, Nick	218
Hutulyk, J.	121
Jakic, Maso	209
Javorski, C.	549
Jenkec, Tony	120
Juretich, Paskoval	104
Jurevir, Luka	733
Jurich, Mike	107
Karoghan, Mike	134
Keyan, John	215
Kinschuk, Harry	122
Klistra, Pete	132
Kobec, Wasyl	261
Konszechin, Nick	424
Kos, Mike	281
Kos, Steve	260
Kostik, Harry	409
Kowal, Sam	329
Kowitsky, Steve	297
Kozak, Joe	135
Kravetsky, Steve	297
Krenbovich, John	185
Kusnik, P.	84
Kuszlir, Mike	256
Kuty, Sam	151
Kuzyk, John	155
Lanoliuk, Steve	118
Lazich, Sam	361
Lazlem, Peter	202

Leski, Mike	333
Luchrosn, George	220
Lukich, W.	795
Lynczyn, Alex	13
Madkovich, Tony	162
Makna, Pete	213
Manopovich, Rade	208
Maraknich, Philip	130
Marich, Tony	267
Markovich, Mike	799
Masadyk, Mike	52
Medich, Bob	312
Medovich, Joseph	183
Mehero, Nick	770
Mesik, John	156
Mezum, Fred	31
Michaelovich, Mike	257
Moar, Dominic	717
Moar, Richard	718
Mossler, Robert	800
Mudry, Joe	321
Murkovtich, Nick	152
Myosk, John	516
Nacinchuck, Gregory	285
Nestoruk, Hercules	530
Niekel, Joe	49
Olbert, John	223
Opola, Mike	485
Osajczuk, John	331
Osiniak, Pete	200
Osmach, Steve	283
Osvilgie, John	192
Oystuyk, Fred	148
Palc, John	27
Palypczuk, Wasyl	140
Pavich, Joe	750
Pavichuck, George	123
Pelich, Dan	337
Perich, Stojan	106
Peskovich, Steve	266
Petresky, Nick	284
Petryk, George	10
Pevich, Ily	263
Polich, Frank	129
Polomchuk, Mike	303
Polonik, Ivan	110
Polovina, Ilia	
Polypovich, Joseph	153
Popil, John	336
Popovich, George	246
Popovich, Nick	164
Porchila, John	738

Poslowski, Bill	273
Potolasky, Cassia	241
Powasczuk, Onufriy	328
Premier, Dan	626
Pros, W.	142
Radnor, Dan	230
Ramps, Frank	271
Rebich, Eade	797
Ridup, Alec	326
Rogousky, John	258
Rosich, Andre	180
Rosich, Nick	295
Rown, Eli	796
Rubich, John	210
Rudyk, Pete	332
Sabich, Nik	711
Schmidt, Peter	344
Shankar, Mike	265
Shapka, Wasyl	
Shewchuk, Bill	227
Shewchuk, John	211
Sikowski, Adam	412
Sitko, Steve	150
Skoreyko, Bert	29
Skypka, Nick	199
Slobodian, John	163
Slonecki, Peter	207
Smikiel, Steve	154
Soich, Joe	357
Srok, J.	113
Stadnichuk, Fred	249
Stanich, Joseph	798
Stonka, Stan	279
Strep, Henry	196
Stricka, Stanko	305
Stuski, Alex	144
Susak, Andy	117
Synazyn, Alex	138
Synecovich, Carl	237
Synigan, John	324
Tarnaveski, Mike	181
Terlik, John	794
Tezechsi, Dan	56
Todoruk, Harry	250
Tonick, Nick	497
Tower, Steve	751
Towir, John	255
Vanco, Joe	301
Viegy, Harry	149
Vimich, Nick	474

Visogoi, Joe	22
Wasylyk, Steve	277
Wistowski, John	517
Wolski, John	619
Woytyla, Joseph	793
Zigich, Mike	362
Zurch, Ignaz	203
Zurich, Mike	288
Zymter, John	44

MORRISSEY

NAME	NUMBER
Ambrester, Phillip	358
Andrechuk, Mike	1
Baby, Tom (Bobby, Tom)	426
Bacha, Joseph	13
Balasz, Gabor	350
Bauer, A.	252
Bednarsky, John	166
Belanko, Sam	345
Belich, Mike	203
Belish, Mike	188
Bertus, Joe	2
Beydo, Mike	316
Biell, Frank	12
Bilos, A.	
Birry, John	201
Bizik, Andrew	16
Bloka, Peter	10
Bonaliuk, Mike	327
Borate, Louis	18
Borsik, Steve	9
Bosak, Albert	17
Boy, George	354
Brauer, G.	
Brink, A.	359
Burke, Steve	334
Calusie, Fedyo	336
Cebar, Lewis	318
Chichy, John	
Chikon, Felix	315
Churla, Mike	
Churla, Peter	
Cohn, George	420
Conjar, Ivan	208
Cook, Louis	347
Corach, Nick	338
Cowch, William	197

Cowch, Mytro	198
Denisker, Anton	195
Deretic, Steve	344
Dobosh, Steve	353
Doskoch, William	207
Dzurgriski, Kasmyr	31
Evasiuk, Steve	333
Fedak, Peter	489
Fikun, Andrew	33
Frade, Alex	34
Fuchak, Peter	332
Gagos, Koskew	309
Galik, Steve	39
Gall, Joseph	43
Gawryluk, Bill	176

Entry gates to compound, Morrissey

Gergas, Frank	
Gillisen, John	
Gorsky, C.	229
Grunthal, Walter	
Gubronsky, John	352
Gulas, Mike	36
Haveruk, Joe	52
Heidner, Albert	
Hein, H.	
Herchuk, William	202
Hlach, Peter	57
Hladysk, John	312
Hlibczuk, Fred	58
Hovanecz, Joe	51
Huelsman, Gregor	362
Hulner, F.	
Hulwa, F.	206
Hrpka, Peter	55
Hrubos, John	47
Hunlin, John	319
Ivanczuk, Mike	59
John, K.	
Jovasevich, Dragon	323
Kaerger, Gustav	948
Katalinick, Mike	335
Kato, Bill	348
Kerczuk, George	
Kessler, H.	
Klingenstein, F.	
Kloth, Gustav	249
Koch, Thomas	958
Kolak, Mike	355
Kolotylo, Mike	180
Konovalchuk, William	79
Kostynuk, Andrew	307
Kovoluk, Mike	66
Kowey, Nick	314
Kozak, Harry	67
Kratkey, Joseph	341
Kronert, Valentin	379
Krusuavich, John	204
Krzuk, John	75
Kubenec, Andrew	76
Kubinec, John	73
Kubos, Steve	77
Kumorak, Joe	
Kupanjar, Steve	326
Lapowsky, Stephen	193

Name	No.
Latka, Steve	85
Lyczynski, John	210
Macenko, Tanasy	199
Machelanko, John	196
Makara, Jacob	92
Malatinka, Joe	93
Manor, Jan	
Marmol, John	99
Marmol, Mike	98
Mart, Frank	96
Marusiak, Vendel	101
Mathias, Fritz	
Mezum, Fred	425
Michaluk, Fred	100
Michelenko, John	196
Mihalsky, John	90
Milder, Marco	189
Mirka, Nick	86
Miservich, Alec	191
Mithas, Fritz	173
Mortiz, Mike	190
Muttobzia, Mike	342
Mychaylyzyn, Mike	308
Myszck, Mike	95
Nikiforuk, Mike	170
Noga, Martin	
Novotni, Andrew	104
Oldorff, Fritz	366
Orlowsky, Sydor	193
Osmanich, Mike	328
Parisha, Frank	112
Pavich, Carl	329
Pavich, Frank	205
Pavich, Matthew	330
Pawlicki, Stanley	
Pawlik, Frank	194
Pechota, Albert	
Petrish, John	119
Petros, Tony	321
Pierko, Toni	117
Pohbralner, Alex	311
Pregler, S.T.	349
Press, Andrew	109
Proskurniak, Prokofyi	178
Rack, Hugo	
Raseluck, Joe	456
Rawluk, Harry	182

Urinak, Tom	148
Urwaslavich, Dragon	323
Vad, Charles	351
Vadakovic, Dan	337
Vatraly, Mike	159
Vehor, Dusan	324
Vollmer, W.J.	
Von Appen, F.	393
Vudo, Joe	153
Vukish, John	183
Wagisch, Petro	305
Wasek, George	
Wasylick, D.	323
Welegan, Mike	
Wolff, G.A.	253
Zagorsky, Mike	209
Zeleznik, Andrew	162
Zubek, John	163

Behind barbed wire, Morrissey

Index